ETHOLOGY

The Biological Study of
Animal Behavior

Rémy Chauvin

Translated by Joyce Diamanti

International Universities Press, Inc.
New York

Dépôt légal.— 1re édition : 3e trimestre 1975
© 1975, Presses Universitaires de France
Tous droits de traduction, de reproduction et d'adaptation réservés pour
tous pays.
Translation, Copyright © 1977, International Universities Press, Inc.

Library of Congress Cataloging in Publication Data

Chauvin, Rémy.
  Ethology.

    Translation of L'Éthologie.
    Bibliography: p.
    1. Animals, Habits and behavior of. I. Title.
QL751.C56713   591.5   76-46818
ISBN 0-8236-1770-X

Manufactured in the United States of America

Contents

PART II
THE PAIR AND THE FAMILY

PART IV

Foreword

Concentrating the essence of ethology, or the science of animal behavior, into some 241 pages presents quite a challenge. In any event, certain choices must be made. Where does the science of behavior begin? Where does it leave off? Whatever boundaries one decides on, they will smack of the arbitrary. For the science of behavior does not, as some believe, consist solely of observations of wild animals in nature, supplemented by a few experiments. It also comprises countless findings made in the laboratory under very rigorous conditions and according to precisely designed strategies. In the field, researchers may err through lack of rigor and succumb to anecdote. In the laboratory, they often go to the other extreme; conditions that are too strictly controlled and an environment that is too poor in stimuli can deprive the results of all ecological meaning and in the end make them uninterpretable. One must, however, take into account the findings of both groups, despite the more or less open war that too often pits one against the other.

In addition, the science of behavior is moving by imperceptible degrees into the field of neurophysiology, that is, there are increasing attempts to interpret behavior in terms of the neuromuscular substratum. Studies in this area are absolutely essential. I state here and now—and will not hesitate to say again—that neurophysiology alone holds the key to behavior. *Only neurophysiologists can explain behavior*; on the other hand, however, *only ethologists can tell what must be explained*. Because of experimental constraints, neurophysiologists are limited to considering only very simple behaviors in very impoverished situations (though this is changing with the development of long-distance monitoring techniques—biotelemetry—and permanent implantation of cerebral electrodes that can be activated by radio). Such experimental situations fall far short of encompassing all behavior; in fact, they frequently comprehend only a fraction thereof, the emphasis being placed more on the physiology of the nerve paths.

We therefore have an excuse for by and large ignoring in this book the neurophysiological study of behavior. Apart from the fact that that field is not our specialty, we are really forced to such a decision because of the abundance of strictly behavioral material that must be presented in so slim a volume.

To begin with, then, we shall discuss the current ethological approach to animals, those marvelous machines that can only be compared to the most complex computers. That comparison is so apt, in fact, that even though it may raise some eyebrows, we shall lead off with a few words about computers. Actually, despite its deficiencies, analogy can be the most useful of reasoning processes. Twenty years ago, this analogical tool was not available to us because computers were too primitive. Nevertheless, we attempted comparisons, which for a time caused us to restrict our study to behaviors that were too simple and even led us to believe that all ethology could be reduced to such elementary components. That was the tragedy of *reductionism*, a kind of naive atomism which greatly impeded the development of ethology. But perhaps we could not have done otherwise.

Be that as it may, the theories that preceded the age of computers are, in our opinion, only of historical interest today. It is sufficient to review them briefly, which we shall do in the introductory chapter.

And then, what kinds of behavior are we going to examine? Since the comparison with computers frees us from inhibiting restraints, we believe that, contrary to the usual approach of books on ethology, the most complex behaviors should be tackled right off. Since animals resemble computers more nearly than anything else, no purpose is served by proceeding as if a computer were a bicycle. It is immensely more complex, but above all *it is totally different*. We must not shrink from stating frankly and fully what needs to be explained; it is not simple—quite the contrary. We believe it would be intellectually misleading to dwell at length on an analysis of elementary movements and displacements, implying that such an analysis suffices to explain everything. We do not intend to fall into that reductionist error. To repeat, nothing is to be gained by taking computers for bicycles; that leads nowhere.

In an ethological approach, two types of behavioral situations immediately come to mind: those in which the animal is confronted with a problem (of predation, construction, orientation, etc.) and those in which the animal interacts either with its mate and offspring or with other members of the group. Thus our investigation of behavior will fall into three subdivisions: (1) the solitary animal, (2) the family, and (3) social relations. In each of these subdivisions, we shall by no means—as we said earlier—ignore laboratory research. Difficult as it is at times, we shall

nevertheless attempt a synthesis between the laboratory and
the field. Finally, we felt it would not be superfluous to
conclude with a few comments on the definition of certain
ethological concepts.

INTRODUCTION:

Historical Survey
of the Development of Ethology

The behavior of animals has always fascinated man, whether he thinks he recognizes himself in them, or whether in the apparent differences of instinct he finds reasons to be proud of being human. From earliest times, those two tendencies can be distinguished: the one to compare animals to man in every respect (with no qualms, for example, about speaking of a moral sense in ants) and the other to regard animals as simple machines having nothing in common with man except a superficial resemblance (if the bitch of Malebranche yelps when one beats it, that is merely the squeaking of ungreased wheels).

Since neither of these attitudes has a scientific basis, one need not tarry long over them. It is interesting to note, however, that they may have carried over into quite modern theories, such as the *mechanist* theory, which tends to view animals only as machines (and rightly so, except that its conception of these machines is far too simple), and other theories that are, supposedly at least, more or less *mentalist* or *vitalist*, which attribute to animals the formation of "mental images," goal-directed behavior, and sometimes even "intelligent" behavior. These latter theories are wrong in employing outmoded and inappropriate philosophical terms, but they are right in pointing out the lacunae in mechanist theories.

Ethology truly came into being toward the end of the nineteenth century when men ceased to believe that they could divine what an animal was thinking and feeling, assuming that it does think and feel. They realized that our only recourse is to *observe and measure what an animal does*. This approach only appears to be modest; objective observation of animals has led to such astounding progress that one now wonders what such a method might yield if applied to man. Too long, in fact, have we been preoccupied exclusively with what he says; it is time to consider what he does.

Another advance came when the concepts of instinct and intelligence were laid aside as inadequate and arbitrary. For years the only concern had been to decide whether a given behavior should be pigeonholed under "instinct" or "intelligence." Eventually it was realized that one could very well study problem solving, rote learning, manipulatory ability, etc., without employing those two controversial terms. They were simply categories established by the Greek philosophers for their own convenience; contemporary ideas about what is convenient or inconvenient, however, differ a great deal from those of Plato and Aristotle.

The founders of scientific observation of behavior were Loeb, Pavlov, and Watson. Although all three were strict mechanists who thought of organisms in terms of the rudimentary machines of their day, they nonetheless succeeded in blazing a vitally important trail. They made the breakthrough from anecdotal ethology and demonstrated the possibilities of objective observation. For example, the study of *tropisms* (a term invented by Loeb to designate the action of "turning toward," from the Greek *tropé*, turn) examines the action of organisms that turn toward a source of stimulation. Purists had refused even to consider active movement toward such a source, deeming it of little interest! All that seems quite naive to us today. Furthermore, as luck would have it, the greatest interest had been in phototropism, which of all forms of tropism is no doubt the most debatable; in many instances, phototropism is unquestionably a pathological phenomenon. On the other hand, orientation toward a chemical source of olfactory or gustatory stimulation (chemotropism) and orientation toward heat and humidity (thermo- and hygrotropism) are much sounder concepts, and biologists still use them today.

Conditioned response, the inspired discovery of Pavlov, has not become obsolete in scientific circles as have certain forms of tropism; however, rather naive application of this concept has caused it to lose favor among ethologists. Conditioning has remained a basic working tool for neurophysiologists, but behavioral scientists know that any attempt to construct the reactions of animals on the basis of such simplistic notions is doomed to failure.

Turning to the modern period, we come to Lorenz and Tinbergen, the founders of the *objectivist school*. As in antiquity, these naturalists observed animals in nature, but they were armed with all the insights of modern scientific thought. They found that the reactions they observed were far removed from those described by the followers of Loeb and Pavlov, nor were they any closer to the behavior of the rat in the maze so dear to Watson and his school. Organisms seemed to function by means of innate releasing mechanisms which enabled them to recognize an appropriate

stimulus as soon as it appeared. All behavior was highly
ritualized, with part of it, however, acquired by experience
and learning. This view of animals was contrary to that of
American psychologists. For them, everything was rooted in
the environment to which the animal adapted according to
the universal laws of learning; for the objectivists, the
innate was at least as important as the acquired.

The fertility of the schools of Loeb, of Pavlov, and of
Lorenz and Tinbergen was extraordinary. The publications
that have appeared under those various banners number in
the thousands. Where do their theories stand today?

It is hard to say, for ethology is evolving so rapidly that
it is difficult to keep track of its successive developments.
Nevertheless, the fact is that objectivism originated more
than three decades ago, and 30 years is considered old for a
scientific theory. The concept of releasers is under heavy
attack. First of all, the particular structures invoked by that
theory do not always trigger a response as automatically
and completely as objectivists would contend; a measure of
acquisition, very early and very rapid at times, is mixed in
with innate mechanisms. In addition, it is hard to comprehend how, on the one hand, animals respond to extremely
simple lures (according to the objectivists, the red belly of a
stickleback is sufficient to trigger combat) yet, on the other
hand, are sensitive to very subtle stimuli or configurations,
sometimes imperceptible to the human eye, which, for
example, enable them to recognize one another individually.

The technique of using lures, ingenious as they may be,
seems crude today. Once initial enthusiasm subsides, the
actions of an animal in the presence of a lure no longer seem
quite comparable to what they are in the presence of a real
counterpart. One should think of an organism as a network

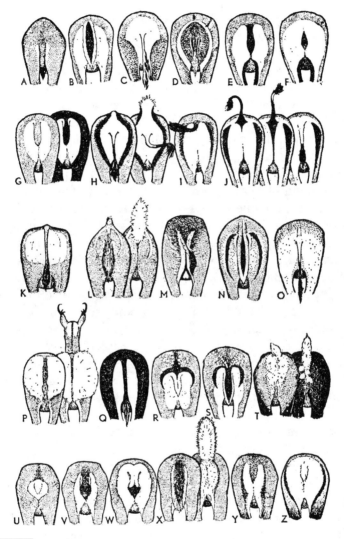

Figure 1. The croup of various ungulates (all reduced to the same size). The markings and positions of the tail constitute widely differing releasers which serve as stimuli in sexual approach or other activities. (After Guthrie, 1971.)

of complex interactions, similar to a chronometer, in which case it is vain to expect that a single factor, or even several, can fully account for anything. In order to analyze the mechanisms involved, which obviously still remains to be done, a better method than that of lures must be found. Some researchers have tried using mechanical models, for example Hunsaker (1962) in connection with several species of *Scleropus* lizards. They bob their heads in response to the presence of an intruder, and it is possible to construct wooden models in which a series of cams produces a variety of bobbing patterns. It seems that in nature females choose a male of the same species by the way in which he bobs his head and they will choose a model with a similar bobbing pattern. But using a more highly developed animal, the chicken, Lill and Wood-Gush (1965) could establish no *simple* correlation between the display of the males and the sexual advances of the females, although it was quite clear that the females discriminated among the males. In the opinion of Cullen (1972), "real behavior involves the perception by a companion of complex, subtle actions which crude dummies have no hope of simulating" (pp. 105-106). On the other hand, the use of implanted electrodes (Delgado, 1967) makes it possible to induce experimentally certain behaviors in living animals with a great deal of precision; however, this remarkable advance in research techniques has not yet borne full fruit.

A more serious concern is that in certain cases *the experiments seem to have been poorly designed*, including some of the most classic experiments such as the one cited earlier on the red belly of the stickleback. In fact, Muckensturm (1969) has found: (1) sticklebacks do react aggressively to an oblong model with a red underside, but they respond equally or even more aggressively to a model with a green or purple abdomen; (2) if the aquarium is illuminated with a red or blue light that suppresses color differences, fighting does not cease; (3) young males who still lack the red that characterizes stickleback sexual display fight one another no less vigorously; (4) the shade of red varies considerably depending on recent experience: for example, a stickleback just beaten in a fight will be a very pale red, whereas the victor will be a brilliant red; logically, a fish with a pale abdomen should avoid one with a bright abdomen, for the vanquished tends to avoid the victor, and that is precisely what happens, contrary to the scenario of the objectivist theory; (5) lastly, to cap it all, individual recognition enters the picture, complicating it to such a point that it no longer bears any resemblance to the classic objectivist schema.

Consequently, as periodically happens in the history of science, the only course that remains is to seek a new approach. As I said earlier, animals resemble nothing so much as computers; we should therefore try to treat them as such, taking our lead from the art of data processing. For example, organisms must perforce contain an innate program, which is none other than the expression of their nervous system and anatomy. It is unlikely, however, that this program is absolutely rigid; that is not even the case with computers, for their program can include instructions to change programs. An attempt should therefore be made to measure the interaction between the program and the environment.

This is not a theory, only a working method. It nevertheless leads us to a marked departure from the usual hypotheses. We shall see this later, particularly in connection with building behavior in the caddis fly (p. 29) and computer simulation of ant construction (p. 26).

We must first of all explore what computers can do and how they function, which can then help us adopt the right approach to living organisms. In the near future, simulation will no doubt become a familiar tool for ethologists. At present, diagrams which are sometimes very complex are used to represent real or supposed interaction between animals (see Fig. 2); arrows of varying thickness give some idea of the probability of action B following action A, rather than action C for example. That is a static picture, however, and when all is said and done we do not really know how well it corresponds to actual interaction.

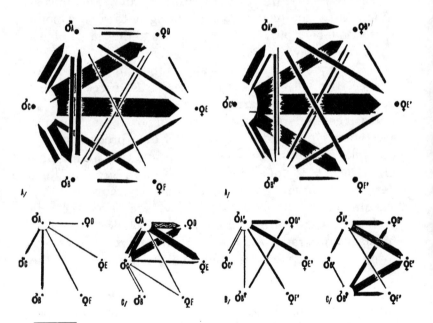

Figure 2. Diagram showing interaction between sticklebacks, the intensity of the interaction being proportional to the thickness of the arrows. Ethologists make great use of such diagrams, the type shown being one of the simplest! (After Muckensturm, 1969).

The computer takes us one step further; it sets that schema in motion, so to speak, by means of its program. And the results are sometimes quite unexpected. Of course a program only shows that reality may conform to that program, not that it necessarily does. Further investigation is required to determine whether the organism does in fact have the same program recorded in its nervous system. On the other hand, if the computer clearly shows that the program does not work, if the results are way off the predictions, then the problem is more serious. Obviously, the ethologist will have to revise the program entirely.

PART I

THE ANIMAL
ACTING INDIVIDUALLY

CHAPTER 1

What a Computer Really Does

The first question that must be answered is whether computers are simply machines for manipulating numbers. It seems that they are now more than that. In any event, they all operate in a similar way. They consist of:

1. One or more "input" mechanisms which transform external symbolic information into an internally usable form (for example, a punched-card reader).

2. One or more "output" mechanisms which transform internal information back into an externally usable form (for example, the device that prints the symbols that come out of a computer).

3. One or more "memories" capable of storing information before, during, and after processing.

4. An "arithmetic unit." Numbers constitute one of the possible forms of information output and it is this unit that produces numerical information. A chess-playing computer, however, does not put out information in this form.

5. A "control unit" which possesses the necessary instructions to activate other units at the desired moment. These instructions are usually very elementary, such as take

---

This chapter is based on the excellent collection of articles edited by Feigenbaum and Feldman (1963).

a symbol out of the memory unit, return this one to the memory unit, shift that one five places to the left, etc.

No one doubts that a living organism possesses all that—and much more.

1. THE BEHAVIORAL MODES OF COMPUTERS

a. *Are computers limited to examining all the possibilities one by one?* If this were so, a computer's behavior could never be called "intelligent"; its only advantage would be greater speed as compared to man. But let us consider the chess-playing machine. There are approximately 10^{120} possible continuations to explore in determining the best alternative move for the pieces on a chessboard, yet in a century there are less than 10^{16} microseconds in which to explore them (Shannon, cited by Newell et al., 1963). It is obviously impossible to construct a machine that can scan them all at every move; the machine must invent a strategy. That is precisely what it does, and that is also why it can be beaten by a man. From this we can leap to the conclusion that an animal living in the very complex conditions of nature could not possibly react by trial and error—it would never have the time. That old hypothesis belongs to the realm of fantasy; it has no validity except under arbitrarily simplified laboratory conditions.

b. *Heuristic programs and algorithms.* Heuristics consist of various methods (strategy, simplification, rule of thumb, etc.) which limit the solutions to be sought in complicated problems. They do not, however, guarantee a solution, nor that one they do propose is optimal; they seek solutions that are merely good enough most of the time.

Algorithms guarantee a solution, provided enough time is devoted to the search. In the case of large computers, algorithms are used for only a limited number of rather simple problems where the machine has time to go through all the hypotheses.

For an example of the heuristic method, let us take the game of chess. One can lay down the following rule: exclude any solution that puts the queen in immediate danger. That is simple and easy to follow; however, it rules out the possibility of finding those brilliant plays that sacrifice the queen if need be. Another very useful heuristic rule is to attack a new problem by the same methods that served to solve an earlier problem of the same type.

c. *The development of a chess-playing machine.*[1] Chess has intrigued many researchers because it is the intellectual game par excellence, of almost infinite complexity and with no element of chance (see Newell et al., 1963). Their first thought was to try to find a technique very different from the way in which man plays, something that might be like the wheel as compared to the human leg; despite completely different mechanisms, they hoped to achieve excellent results. So far, however, there is nothing of the sort in sight. Playing chess, it seems, requires a very complicated program. The need for such complexity was only gradually perceived.

The development of chess programs began with *Shannon's proposal*, first set forth in a paper in 1949. Shannon proceeded from the idea that every move eventually results in a win, a loss, or a draw; a diagram of the game would look like a tree with numerous branches corresponding to alternative moves and the problem would be to move in the direction of ultimate maximum gain. This is achieved by a

---

[1] I use the term machine for convenience only; it is actually a matter of chess-playing programs of varying complexity which are put into *an ordinary computer*.

procedure called minimaxing in the theory of games: a player assumes that alternative moves are made, sees what can happen, and by reasoning backwards chooses the best alternative. Thus Shannon's proposal consisted of: (1) considering alternatives, (2) analyzing the consequences, and (3) choosing the best alternative. The analysis is the hardest part. It must be divided into three steps: (a) explore the consequences of each alternative to a certain depth, two moves deep for example, because it is not feasible to go further; (b) at the end of each exploration evaluate the position of the pieces; and (c) combine those evaluations to form an over-all value for each alternative. The machine would then select the move with the highest value. Shannon did not, however, design a specific program.

Turing's Program: Turing (1950) designed a program along the lines laid down by Shannon with a few modifications, especially as regards evaluation. He took into account the respective value of the pieces, which is obviously not equal. Turing's machine was put into operation, played rather badly, but nevertheless managed to beat a poor player.

The Los Alamos Program. This program (Kister et al., 1957) is again similar to Shannon's, but with a chessboard of only 36 squares instead of 64. It succeeded in beating a mediocre player. The machine required an average of 12 minutes to make a play, examining about 30,000 positions per move; a human player considers far fewer than 100 positions.

Bernstein's Program. Bernstein constructed a chess-playing program for the IBM 704 (Bernstein and Roberts, 1958; Bernstein et al., 1958). In this case the program is heuristic and the machine considers only some of the alternatives by adhering to a few safety rules: ensure the king's safety, defend one's own men, attack the opponent's men, etc. Only seven alternatives are taken into consideration. None of these programs is capable of beating a good player.

d. *The machine that learns to play checkers.* The game of checkers is of special interest as compared to chess because it is simple and lends itself better to the study of machine learning (Samuel, 1963). Once again, however, not all continuations can be taken into consideration, the reason being the same as in chess, i.e., there are just too many possible moves of the pieces.

On the whole, the tactics resemble those of Shannon rather closely. A move is regarded as a move initiated by the machine + the countermove of the opponent, and the machine explores the consequences three moves ahead (more would not be practicable). Each move is assigned a coefficient according to its subsequent success in terms of the criteria adopted and the machine selects the alternative with the highest coefficient. The criteria are based on the inability of the opponent to move and the number of the opponent's pieces (which the machine tries to reduce).

A vital part of the machine is its memory. It cannot, however, register all the consequences of a move; it must do with recording the continuations for only three moves. It must be added, though, that this is very effective, for when the machine advances a piece, it calculates the consequences of this move three moves ahead. It follows that if by chance one of the anticipated moves has been registered previously, then with the aid of its memory the machine can follow the continuations six moves ahead. The program must also include directions that lead toward winning rather than merely moving the pieces about.

The memory keeps track of the number of pieces on the

board, the presence or absence of a piece advantage, the side possessing this advantage, the presence or absence of kings, and whose turn it is. Board positions are recorded and the memory explored in the order of a decreasing number of pieces: first, positions with 12 pieces to a side, then 11, 10, etc. After a time the memory may become cluttered by a long series of games. Then "forgetting" is introduced by taking into consideration how many times each previously recorded board position has been used by the machine. When it has not been used for a given number of moves it is eliminated.

With this program (presented here in simplified form) the machine actually learns by playing, extremely badly at first, and then, after 20 or 30 games, it is classed by checkers experts as "better than average." We should like to point out, however, that the programmers could not understand and hence could not program some of the strategies used by checkers champions.

e. *The problem of perception: Recognition of patterns or letters.* One of the first efforts in the field of mechanical perception was a program for recognizing Morse code signals and transcribing them directly into letters (see Selfridge and Neisser, 1963). That would not be complicated if human operators sent ideal Morse, that is, if the dots and dashes respectively were always of fixed lengths and if the spaces separating marks or letters were also always uniform. But that is far from being the case: some dots are longer than some dashes and spaces are sometimes longer between those marks than between letters. It is quite remarkable that this gives an experienced human receiver little trouble. He even gets to the point where he no longer perceives the dots and dashes as such when he hears them; he perceives the letters directly. A machine, then, must first learn to identify letter spacing by the heuristic method: out of six consecutive spaces, the shortest is accepted as a mark space (separating the marks within a letter) and the longest as a character space (separating letters). The effectiveness of this rule can be demonstrated by experiment; in practice, it proves to be in error only once in 10,000 times.

The dots and dashes received by the machine are translated into a sequence of numbers measuring their duration, which also makes it possible to measure their spacing. The machine's rate of error is only slightly higher than that of an experienced human operator.

Constructing a machine that recognizes hand-printed letters is more complicated. Letters can be printed in a great many different ways and they may also vary in size. Notwithstanding, a human reader automatically recognizes them instantly and unerringly. In machine recognition, the letter is first projected onto a matrix of fixed size made up of a given number of contiguous photoelectric cells, some of which will be blacked out by the letter. The machine scans the bank of photocells one by one and translates the whole into binary terms: 1 for the blacked-out photocells, 0 for the others. The letter is then translated into a long sequence of zeros and ones which follows a pattern typical of the letter in question. The machine must then evaluate the similarity of this pattern to a certain ideal template.

But things are not so simple. Even if the pattern matched, a slight change in slant or size could completely destroy the comparison. Nevertheless, a given letter can be defined by certain characteristic features. For example, A tends to be narrower at the top than at the bottom, more or less concave at the base, etc. Here is part of a decision-making sequence by a machine, the task being to distinguish the letters A, H, V, and Y:

Concavity above?
Yes   No . . . . A

Crossbar?
No   Yes . . . . H

Vertical line?
No V
Yes . . . . Y

In certain programs it is truly remarkable how after a time the machine shows all the characteristics of associative learning. For example, generalization of stimulus and response (Feigenbaum, 1963): if X and X^1 are similar stimuli and Y is the response that should be given to X, then by stimulus generalization it will become possible for Y to be given as a response to X^1 also. Generalization is a result of the structure of the discrimination net. For similar stimuli in the machine's memory are those sorted to the same terminal and similar responses are those stored in the same area of the net.

One also finds oscillation and retroactive inhibition. These phenomena are consequences of the increased complexity of the net. Information sufficient for orientation within the net at one time may no longer be so when the net has expanded through the introduction of new items. Both phenomena have the same cause. This relationship between oscillation and retroactive inhibition has not been clearly recognized by psychologists.

Finally, forgetting also occurs in machines. It does not result, however, from the destruction of information, which is in fact integrally preserved, but rather from the accumulation of information which eventually makes it difficult or impossible to retrieve a given item within a given period of time. This aspect of forgetting is another mechanism that has not been considered by psychologists.

f. *The machine that shows "mood" or "motivation."* Gullahorn and Gullahorn (1963) developed a computer program to manifest certain kinds of human behavior. It is true that this is difficult to feed into a machine, unless one accepts, as the Gullahorns do, the "social exchange" (rather unsophisticated) of Homans (1961), which he set forth in a whole series of propositions, using as an illustration a group of government employees who have the same rank and work in the same office, but one of whom is more competent and the others seek his assistance. The great advantage of these propositions is that they can be easily formalized and put into a computer.

Proposition 1: If in the recent past a particular stimulus situation has been the occasion for a reward, then the more similar another situation is to that earlier one, the more likely it will be to elicit the same or similar activity.

Proposition 2: For a certain period of time, the more often one man's activity rewards the activity of another, the more often the other will emit the activity.

Proposition 3: The more valuable a man finds a unit of activity given to him by another, the more often he will emit the activity rewarded by the activity of the other.

Proposition 4: The more frequently a man has received a rewarding activity from another in the past, the less valuable each unit of that activity will be to him in the future.

One can see how this new behaviorism might lead to a precise study of social reactions, to veritable machine simulation. But research is not yet far enough advanced in that direction.

If perhaps I have dwelt too long on ideas that will seem commonplace to some, it is because *we do not fully understand what we are imitating or reproducing*. This is a familiar working concept for engineers and it laid the foundation for bionics. As complicated as the program of a chessplaying machine may be, that of an ant or bee must be much more so. At the very least, then, any attempt at explanation must be more complicated than the program of such a machine! But many biologists are convinced of that only in theory and not in practice.

2. ETHOLOGICAL DESCRIPTIONISM

Let us go back now to what ethologists do, or at least to what most of them do. Many years ago Lorenz, following a number of illustrious predecessors (and he is the first to acknowledge them), quite rightly called attention to the need for exact and thorough description of behavior. The worst mistake of all when studying an animal is not to look at it, and that is precisely what the rat-in-the-maze fanatics did for so many years, without exception, and what some physiologists are still doing today.

One must, therefore, look at the animal closely and thoroughly. But what is there to look at? From the standpoint of systematics, I maintain that a behavioral peculiarity—for instance, the way in which herbivores scratch themselves—is just as informative as a morphological peculiarity. And when the latter is ambiguous, sometimes reference to the former can definitively settle a question of classification. But is ethology restricted to detailed description of every little movement an animal makes? I, for one, do not think so, and I offer the following observations in support of my position.

1. An early experiment showed us that rats can negotiate a maze while swimming, even though they learned it

on foot. It is obvious that the gestural sequences and the whole coordination of movements, not to mention the stimuli, differ very much in the two situations. Nevertheless, that does not appear to trouble the animal.

2. In the case of the checkers-playing machine, it is of course understood that the machine does not touch the pieces; it merely indicates the moves to be made by means of a conventional code and the operator places the checkers in accordance with the machine's directions. Actually, there is nothing to prevent us from equipping the machine with articulated arms so that it could perform that task itself. Clearly, however, *the important thing is not to describe the movements of such arms, but to describe the program that determines those movements.* Mechanical arms add nothing significant to the machine's directions for the moves.

3. We shall see later, in the discussion of the building behavior of Trichoptera, that an individual caddis fly larva repairs its case in all sorts of different ways (p. 33).

There are two questions: (1) What do animals do? (2) How do they do it? Are these two questions dependent on one another and, if so, to what extent? Is one as important as the other? If one could prove that *an animal has an alternate heuristic*, that is, that it can try different solutions from its program when it is unable to use the normal solution, then the question of "how" would be less important.

Does completion of the program hinge upon the movements? That point is uncertain; unfortunately very few experiments are available that have probed deep enough to permit an analysis. I have found almost no examples that are sufficiently conclusive except in the case of the caddis fly larva and the bagworm (*Psyche viciella*), both studied by the Polish school (Dembowski, 1933; Staropolska and Dembowski, 1958; Gromysz, 1960). I think that equally convincing examples must exist among birds, especially weaverbirds who perform complicated repairs on their nests, as Crook (1964) has shown. His work, however, is in quite a different vein from that of the Polish school. The repair techniques that he inventories do not take individual variability into consideration; he does not introduce obstacles to construction that would oblige the animal to take quite another tack in order to carry out its program. I have no doubt, though, that it could succeed.

The importance of the program. A conclusion emerges from all these studies that has met with opposition in some quarters: isolated gestures are only of relative importance, the main thing being that the program "tries to come through." Not that such gestures are insignificant nor that a careful description of them cannot lead to interesting results; but those results will be in the sphere of comparative anatomy and embryology. Study of the program, however, belongs more to the science of behavior.

The importance attributed to movements can be seen in the lengths researchers go to in order to describe them coherently and feed them into the hungry maw of the mathematical mill (for ethologists and psychologists, feeling excluded from the temple of science, hope to gain entry by way of mathematical formulation). In order to do this, they are forced to break up sequences into tiny "bits"—that is the term they use—and also to select their subjects.

I am not opposed to selection of subjects, provided it is clearly stated what percentage of them "react as they should." For if this percentage is too low, one may well question the importance and significance of the results obtained. And even if it is high, one must wonder what the behavior of the remainder consists of, those scorned and secret dregs that are filtered through the equations.

Changing methods. It is incumbent on anyone studying programs to try to bring to light all their possibilities. Only very rarely has this been accomplished. Natural programs are much more flexible than reductionist experimenters imagine. As we have seen, the work of Dembowski and the Polish school is an excellent example of such research. The publications of Muckensturm (1965a, 1965b) on the way in which the stickleback removes obstacles placed around its nest, those of Darchen (1959) on the techniques used by bees in nest reconstruction, and my own (Chauvin, 1958, 1959, 1960, 1965, 1970, 1971) on construction in ants show other applications of that particular research method.

The best area of study is the nest. Long ago I stressed the importance of animal building behavior (Chauvin, 1956). It is in construction that the complexity of a physiological program is most fully expressed. It is there that experimental interventions elicit the most immediate, most complex, and at times most astounding responses. If we are to bring all this to light, we must not be content with merely describing in detail the usual techniques of construction, which ethologists have already done with great precision. Like Dembowski and Gromysz, we must turn to reconstruction after the nest has been damaged or destroyed, we must introduce obstacles, as Darchen and I have done, for that is the only way in which the enormous resources of a program can be revealed. It is clear that *the focus of the naturalist's concern is making a marked shift toward what animals do, rather than remaining fixed exclusively on how they do it, for in point of fact they can do it in a hundred different ways.* That is the astonishing conclusion that can be drawn from research such as that of Dembowski and Gromysz.

Once one thinks one understands a program, there is only one possible proof to administer, indirect though it may be: computer simulation. Gallais-Hamonno and I experimented with simulation of mound-building in ants (Gallais-Hamonno and Chauvin, 1972). Of course success of a simulation merely shows that things may happen in a certain way and not that they necessarily do. But failure of a simulation plainly shows that the subject's program has not been understood.

We should not forget that even a very simple computer program is nevertheless *infinitely* more complex than the simplistic schemata used by ethologists to explain behavior. Simulation is altogether different from explaining a simple gesture in terms of basic types of neuromuscular coordination; rather, it is a matter of forming a picture of the behavior as a whole and the computer's telling us whether or not the model works according to our predictions. In the case of the mound-building ants, for example, the program of each theoretical "ant" contained the following:

1. the order to go and seek twigs at random, but moving away from the center of the plane;

2. the order to carry twigs for brief and variable periods of time;

3. a directional instruction: to carry toward the center of the plane (not necessarily to the center, but in the direction of the center);

4. a prohibition: not to put more than 10 twigs on top of one another at the same point (because of balance).

Many other instructions were devised, but only with these did the graph traced by the computer take the shape of a mound. I should like to point out that the first order is already complex, and it is generally at that level that etho-logical analysis stops. Behavior simulation, however, can be carried out only at a much higher level.

CHAPTER 2

Construction

1. BUILDING BEHAVIOR OF CADDIS FLY LARVAE

Larvae of Trichoptera (the generic name of caddis flies), which must number 10,000 species, can be divided into five classes according to their habits (Ross, 1964). In the first, which comprises four families, the larva constructs a fixed abode or net, attached to some object on the stream bottom. In the second group, the larva is mobile, attached to the bottom only by a silken filament. In a third group, the larva conceals itself in a portable tube whose front and rear ends are open. Another group encompasses the common caddis flies; the tube is closed at the rear and the larva protrudes from the front end of the casing, which may assume all sorts of shapes and may be composed of or covered with all kinds of materials.

Caddis fly larvae employ several methods of construction (Hanna, 1961):

1. the tunnel method in which the larva gathers a pile of leaves and crawls under it;

2. the dorsal plate, ventral plate, and side wall method in which the animal builds those various elements of its case in succession;

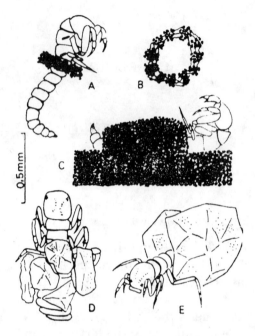

Figure 3. Three types of casing construction of *Molanna angustata* larvae. A and B: abdominal girdle; C: tunnel; D and E: examples of crude shelters. (After Denis, 1966-1967.)

3. the girdle method and T method: having gathered a pile of assorted material, the larva fashions a sort of girdle from it or binds it with silk in the shape of a T;

4. the burrowing method, in which the larva buries itself in the ground up to the thorax and then merely constructs a ring around its thorax.

a. *Selection of materials.* According to Hanna, quality, size, and shape constitute the three important factors in selection of materials. Many species choose the same material throughout the entire larval stage. There are species, however, in which the larvae prefer one type of

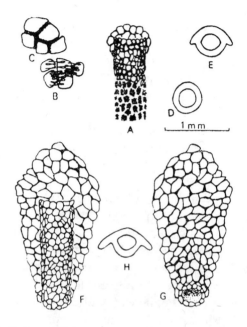

Figure 4. Casings of *Molanna angustata* larvae. ABCDE: first casings built by young larvae; D and E: anterior and posterior cross-sections. FGH: typical early casing; F: ventral view; G: dorsal view; H: cross-section. (After Denis, 1966-1967.)

material when they are young and another type when they are older.

In *Molanna angustata*, the larva selects particles according to their size, surface characteristics, shape, and weight. Size is of prime importance, before weight and texture. The *Molanna* larva prefers light weight, flat fragments with a smooth surface, avoiding those that are heavy, thick, or rough-textured. The time taken to select material may vary from a few seconds to a minute, longer times not necessarily being related to the scarcity of material. The *Molanna* larva selects material to its liking, for example, picking out a fragment of eggshell that is mixed with bits of brick, even though the preferred material may constitute only a small part of the whole. There is *recognition of shape*, which can be proved by providing the larva with bits of eggshell that vary in shape—round and square—but are comparable in size. In that case, the roof of the casing will be built almost exclusively of discoid pieces. Similarly, if the larva is offered eggshell fragments of two sizes, the roof will be made of large pieces and the floor of small ones.

A very interesting experiment consists in giving larvae of *Neuronia postica* both pine needles and tiny pebbles the size of grains of sand. Normally these larvae do not utilize such material; they prefer oblate vegetable debris. If they are given only sand, some may obstinately refuse to build at all. If, however, there are also some woody bits, such as pine needles, then the larvae will bind a few of them together with silk, placing some twigs parallel and others more or less perpendicular. The larvae then glue grains of sand underneath the parallel twigs so that the sand, stuck to the silk, hangs down and forms a sort of semicylinder, which is completed by gluing sand along the sides; then the upper portion is built by gluing grains of sand on top. Finally the twigs, being of no further use, are removed. Thus the larvae use *a kind of scaffolding which is later dismantled*. It should be added that if a larva is given only twigs, it never builds a casing and all its usual construction techniques are markedly changed. Finally, *learning seems unnecessary or at least extremely rapid*. In the very earliest larval stages, *Neuronia postica* are already building typical casings; however, some writers do not rule out the possibility that techniques are perfected somewhat as the larvae grow older.

b. *Construction*. If a larva is removed from its casing, it will wander blindly for a moment or two and then burrow rapidly into the ground headfirst. Thrashing its abdomen wildly, the larva throws grains of sand up over its body toward its head, as if trying to hide itself as quickly as possible. Only later does construction truly begin. Dembowski (1933) distinguishes several phases: (1) selecting particles, which are gathered in one by one with the forelegs and probed by the buccal parts; (2) grasping these particles: having reached out to take hold of a particle, the larva grasps it firmly and then pulls back into the tube under construction; (3) placing the particles in the web of silk that forms the framework of the edifice. On the average, the first phase lasts 9 seconds, the second 10 seconds, and the third 30 seconds. But these times should not be thought of as fixed; on the contrary, they are extremely variable. For example, a larva may seize a grain of sand and place it immediately, or it may discard five or six particles before placing one. When the tube has reached its definitive size, the larva turns around and removes the posterior portion, which it then proceeds to reconstruct. After that, there comes a rest period. Finally, the larva lines the inside of the casing with silk. Only later will it undertake to build flanges by leaning halfway out of its shelter.

c. *Behavioral variability and building repairs*. When selected portions of the casing are experimentally removed, it is most interesting to note the repairs the larva makes to the tube and, in some instances, to the flanges. At first, active repairs are confined almost exclusively to the anterior portion (tube or flanges). Repairs to the posterior portion are carried out much more slowly and laboriously; sometimes they are not made at all, the experimenter's holes merely being plugged up with silk. For example, if excisions

of various types are made in the front third, they are
repaired immediately. If the excisions involve the front two
thirds, however, the larva constructs a new casing, using
the remaining rear third as a support to begin with, but dis-
carding it later. A small round hole made by the
experimenter in the middle of the casing will be repaired,
but in an unusual way: the larva first festoons it with silk
and then sticks mineral debris to it in a very irregular
arrangement.

A most astonishing example of adaptation was discovered by Dembowski. After having placed a larva in its
casing in a laboratory dish which contained no building
material, he removed part of the roof of the casing. He later
found that the missing portion was regenerating, being
covered over with mineral debris taken by the larva from
other parts of its casing!

Larvae may also adopt artificial casings and provide
them with the flanges they lack. If there is not enough
building material (which can happen only in the laboratory, not in nature), larvae will use the walls of the recep-
tacle to make up for the scarcity of material, in which case
the house will not be mobile. Dembowski, however, witnessed an extraordinary phenomenon: after a time he
provided such larvae with additional material, whereupon
they completed a more normal construction and then *cut
the mooring lines and moved the whole edifice*!

As Denis (1966-1967) points out, however, "the behavior of any one individual is never exactly like that of any
of its congeners." Sequences are exchanged, abbreviated, or
omitted, for example, in reconstruction of a girdle casing.
In the course of reconstruction, moreover, a larva that
proceeds to build afresh after having been removed from its
casing will do so in a way "that may be identical to or

different from the way it chose the first time, and if animals are repeatedly removed from the houses they are rebuilding, they may use a series of different methods to reconstruct the framework of their casings."

d. *The stimuli involved and the role of the nervous system.* In a study of *Silo pallipes*, Hansell (1968) describes a sequence of characteristic movements involved in selecting a particle and fitting it into the casing: (1) the "scratch" test, in which the larva rakes in particles that interest it with its pro- and mesothoracic legs; (2) the "handle" test, in which the larva picks up a particle, holds it underneath itself, and manipulates it with all its legs and its mouth parts; and (3) the "fit" test, in which the larva, still holding the particle in its legs, turns over onto its back and fits the particle into the roof of the casing. Hansell has demonstrated that manipulation, which often results in rejection of a particle, corresponds to a tactile reading of the size of the particle. What is most surprising is that the amputation of a pair of legs—no matter which pair—does not alter the phenomenon at all. There must therefore be an astonishing degree of flexibility, and perhaps the program for manipulating and evaluating particles is not dependent on any specific movement. Similar results have been noted after amputating the legs of other species of Trichoptera.

2. Building Behavior of Bees

Observers have always been struck by the beauty and symmetry of structures built by bees. Nevertheless, since Réaumur and François Huber, few biologists have chosen them as subjects for their experiments. Darchen (1959), however, has reported a great many new observations on the building techniques and repair capabilities of worker bees after experimental damage.

Above all, it must be remembered that construction is a social phenomenon; solitary bees or groups that are too small do not build. The task requires a sufficient number of workers and, especially, the presence of a queen; only colonies of several thousand bees will undertake construction in the absence of a queen (Darchen). Furthermore, the mere capping of a cell requires the intervention of at least 300 workers.

a. *Critical zones.* To understand Darchen's work fully, one must realize that in the little ellipsoid combs that bees build first, not all areas are the same. It is in the zones of expansion, that is, throughout the pendent lower extremity and especially along its edges, that experimental intervention has the most conspicuous consequences. The least obstacle, for example a matchstick fragment inserted in the edge, causes a "hernia" to appear over the obstacle, which will later be reabsorbed below. On the other hand, the same bit of matchstick stuck into the edge of the upper portion, not far from where the comb is attached to the wood, produces nothing worthy of note.

A particularly interesting experiment consists of placing a thin metal blade about half a centimeter wide into the edge of the comb; this brings construction to a complete halt on that side so that the comb becomes lopsided. But if the blade is perforated with holes of the right diameter and sufficiently close together, the metal blade will little by little be covered with wax and swallowed up in the construction. If little daubs of wax are stuck here and there on a nonperforated blade, to serve as cues to the worker bees, that is not enough to make them resume construction; the holes are absolutely essential. Also, from time to time one sees worker bees put a leg through these holes or straddle the blade, clinging to their fellows on the other side. Moreover, wherever wax walls are being built, one always finds long chains of bees clinging around the work who seem to be completely at rest (waxer bee clusters). This is what leads Darchen to believe that the crucial factor that regulates building activity must lie in subtle changes in the muscular pull exerted by the legs, which must vary as the work progresses or is altered experimentally. In a way, it would really amount to a "modulated tactile language," the rules of which are not yet known.

b. *Maintaining a parallel plane.* We know that the face of a comb forms a parallel plane; but upon closer inspection we learn that the bees keep it even only as far as possible and do not conform to parallelism at all costs. To keep a comb parallel, however, they are capable of a series of complex adjustments. For example, if a comb is bent back to form a blunt angle, the walls of the cells farthest from the apex of the angle will be extended, while those of the nearest cells will be trimmed back. Since in places this results in cells that are too shallow, the bees may later move the rear walls back in order to return them to normal size.

Darchen, however, devised a more elegant method to bring to light the maintenance of parallelism. It consists in inserting a little perpendicular blade of wax between two rows. In a very short time the waxer bees twist it and work its faces, drawing them out so as to make them parallel to the comb. This is accomplished by chains of waxer bees pulling in opposite directions on the two sides of the strip of wax, following a complex procedure that has not, however, been fully clarified.

c. *Adaptive capabilities of bees.* We have seen that

bees can readily solve rather complicated problems, albeit in their own way. They can also adapt to very unusual circumstances. For instance, if the combs of a hive are placed horizontally rather than vertically, this total reversal of the facts of the problem *provokes no serious disruption*. The queen continues to deposit her eggs, and the workers their nectar, on the upper and lower sides of the comb without seeming to be troubled in the least. Only the orientation of the cells is eventually modified, so that the long axis of the cells on the periphery points up and out on the upper face. In normal combs, in a vertical position, the long axis of the cells slants up on both faces.

Bees are capable of a host of alterations, such as those to keep the combs parallel, for example, and they can also make all kinds of detailed repairs and eliminate many obstacles to their work. For instance, it has long been debated whether the solitary bee called the mason bee (*Chalicodoma*) can repair the bottom of its mud nest when it sees the nectar and pollen it has deposited there escaping through a hole made by the experimenter. It does seem that the bee can do so provided that the experimental intervention is made at a not too advanced stage of the work, for example, during the phase of stocking the nest. If the bee is in the final stage of capping it, then that is the end of construction for the mason bee and resumption of building is out of the question, even if a gaping hole is opened in the bottom of the nest.

With the honey bee, however, it is quite another matter, as Darchen has clearly demonstrated. Holes made in the end or walls of cells are skillfully filled in. Needles implanted endwise in the cells are extracted and thrown out; those embedded in the walls crosswise, tangent to the surface of the comb, are removed by cutting away the partitions to the exact extent necessary for removal, after which the partitions are immediately repaired.

3. Building Behavior of Ants

The mounds of twigs built by red forest ants (*Formica rufa*) are sometimes enormous, measuring more than a cubic meter. Though the surface of the mound appears to be made of twigs assembled haphazardly, it is actually under continual supervision and maintenance. A tiny hollow only half a centimeter deep will be quickly repaired and covered over. A dead leaf that falls on the mound will be removed or, if it is too heavy, covered up with twigs. Bits of wood implanted in the mound will be dug out, even though they may be as big as a pencil. On the other hand, if receptacles of varying sizes are placed on the mound open end up, they will very rapidly be filled to the brim with twigs—these ants have "a horror of holes." This can degenerate into a real mania: the ants can be induced to fill receptacles of several liters whose sides extend well above the surface of the dome. Similarly, they can be made to remove a protuberance caused by the experimenter placing a handful of twigs on the surface. Thus these ants are capable of working not only by adding to but also by subtracting from their mound.

In another connection, if one places an X-shaped partition across the top of the mound, after a few days one of the quadrants will contain far more twigs than the others. It is not, however, always the same quadrant. If one removes the partition and then replaces it once the surface has been equalized, one finds that after a few days it may well be one of the other quadrants that contains the abundance of twigs. It seems that the ants prospect "twig fields" in a certain direction and then bring their booty back to the mound in almost a straight line (observation of the individual paths of several hundred workers shows that to be the actual case). Normally, the effect of this unbalanced supply, greater input from one side than from

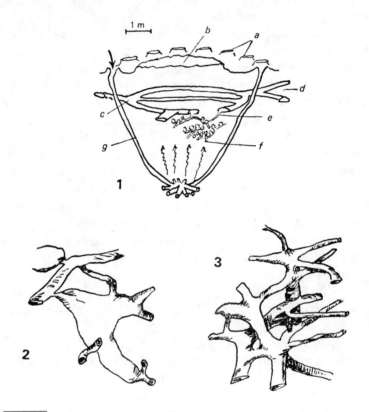

Figure 5. Diagram of the nest of *Alta sexdens*, an ant that grows fungus. 1, a: craters; b: excavated disc; c: circular tunnel; d: entrance tunnel; e: descent tunnels with fungus chambers; f,g: side tunnels; wavy arrows show probable air circulation. 2: a "barracks" consisting of a large bulge in the tunnel network. 3: a large "interchange." (After Sudd, 1966.)

another, is corrected by these ants' "horror of holes" and inequalities are evened out very quickly—unless, of course, a partition has been interposed.

The sense of unevenness is so acute that some furrows only 2 millimeters deep scratched in a board are filled in with very fine twigs. On the other hand, the type of building material the ants bring back to the anthill is not rigidly determined: by denying them any sort of twigs in the laboratory, I have induced these ants to build a mound of pebbles and even steel pins.

Thus the relatively simple building principles of these mound-building ants are apparent. As Rabaud (1929) had seen so clearly, there is no fixed plan that the workers are conscious of and the lack of coordination is *almost* total; but there is a sort of "external coordination," a set of very simple reactions that almost always lead toward achievement of the common task. This is true of the aversion to holes, the blind impulse to gather (but since these are forest ants, the probability is high that they will find twigs and hence their nest will be of twigs), the tendency to pile twigs as high as possible (hence in the end the nest will be mound-shaped), and so on.

4. BUILDING BEHAVIOR OF WASPS

The well-known work of Deleurance (1957) is an excellent example of classic descriptive ethology. He provided remarkably precise descriptions of the postures and movements of wasps while they construct their nests, but he did not explore the ultimate possibilities of their building instinct. Vuillaume (unpublished studies), however, had shown that their capacities are very great. For example, underground wasps can rebury their nest when it has been

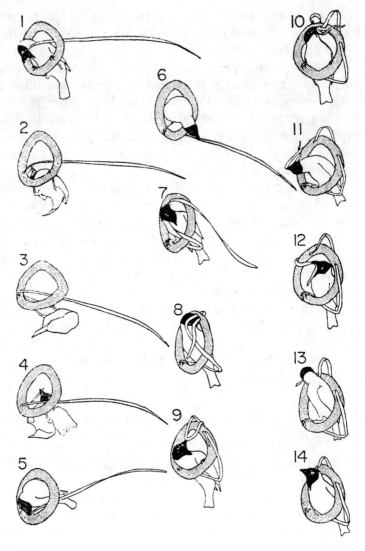

Figure 6. A weaverbird fixing a blade of grass into the ring it has constructed. (After Collias and Collias, 1962.)

placed above ground by calling on behavior mechanisms that would not normally be suspected.

5. Birds' Nests

a. *The weaverbird.* The study of Collias and Collias (1962) on the weaverbirds (*Ploceus cucullatus*) clearly shows the astonishing complexity of these birds that actually weave their nests but will not decline to use wide-mesh screen provided by the experimenter. Study of partial or total destruction of weaverbird nests has begun,

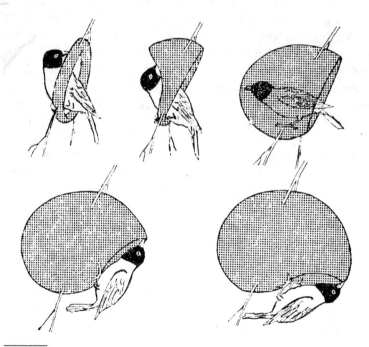

Figure 7. The beginning and finishing stages of nest building in the weaver-bird. (After Collias and Collias, 1962.)

although it is much less advanced than the research just reviewed on caddis fly casings. Experiments show that weaverbirds are capable of complex repairs; therefore, it is not, as had been thought, "each successive act of nest building that of itself produces the stimuli necessary for its own completion and commencement of the next stage" (Collias and Collias, 1962). Similarly, although special problems, such as being supplied with grass that is too short, bring construction to a halt in most cases, there are nevertheless some birds that will complete their nests in such circumstances.

b. *The "thermometer bird."* The mallee fowl (*Leipoa ocellata*), which inhabits a large area of southern Australia, lays its eggs in a complex incubator of its own making. First the ground is excavated and then the hole is filled with moist vegetable matter that will produce heat as it ferments. Finally the bird heaps up earth or sand which will, on the one hand, store solar heat and, on the other hand, prevent dissipation of the heat from fermentation. Several times a day the bird plunges its beak or head into the mound, in all probability to take its temperature. If the nest is too warm, the mallee male will scoop out sand before sunrise, throw it in the air to cool it, and then reconstruct the hillock. If the nest is too cool, the bird will spread out the sand from the top of the mound in the sun, heaping it back up only after the sand is hot. A related species knows how to utilize volcanic heat. At Savo, in the Solomon Islands, there are two sandy areas pocked with hot fumaroles; large numbers of birds congregate there and simply scoop out little cups in the sand where they deposit their eggs. On the island of Niaufou they even nest in the warm ashes within the crater of a volcano.

The hillocks of the mallee fowl may reach an enormous

size, over 10 meters in diameter and more than 4 meters high. Regulating the temperature of thick layers of decaying vegetable matter is obviously not so simple. For instance, the oldest and deepest layers may have ceased to ferment and produce heat, when the top layers are, on the contrary, at the peak of fermentation. The oldest eggs, nearing the end of incubation, are found in the lower layers and the freshly laid eggs in the upper layers. When a bird is gathering vegetable matter, it works slowly if the weather is dry and will not close the incubator until rain falls (quite rare in this region), thus wetting the vegetable layers and initiating fermentation. All this is extremely complicated and it is not surprising that Frith (1962) found that young birds are not very successful at regulating the temperature of the incubator; they need several years of experience to achieve complete success.

Their thermostatic control is very precise and temperature recordings show that they keep the nest at 92° F. in the egg chamber. When the incubator tends to heat up or cool off, the recording instruments prove the effectiveness of the bird's interventions. Frith once built an incubator himself to test the effect of construction on temperature control and he was assisted in this task by a mallee fowl that came every day to note the progress of his work and even alter it to suit itself!

Experimental modification of the nest temperature by means of a resistor brought about curious changes in the bird's behavior. Frith began by heating the nest in the spring, a relatively cool season when the heat mainly derives from fermentation of the vegetable matter; the bird greatly increased the number of vents in the incubator in order to cool it off. When the nest was heated in the summer, however, a season when the heat normally comes from the sun, the bird no longer seemed equipped to control the excess heat from within; instead of increasing the number of inspection holes and vents in the nest, as would be logical, the bird heaped up more sand on it. In the fall, on the contrary, the nest is normally opened up during mid-day in order to trap the heat of the sun, but when it was artificially overheated, the bird left off opening up the nest once it discovered the interior was warm. Therefore the mallee fowl can adapt its behavior to the circumstances only to a certain extent. It should be added that when the bird detects temperatures that should not normally be en-countered, it pokes its beak into the mound repeatedly and shows signs of doubt about what should be done. One part-ner may even proceed to undo what the other has done, but it is usually the methods of the male, the principal builder, that prevail (Frith, 1962).

c. *The bowerbird.* The bowerbirds of Australia (*Ptilonorhynchus violaceus*, for example) build arbors fash-ioned of two rows of dry boughs curving inward and at the entrance they heap up innumerable shiny or bright-colored objects in a display that travelers have always found so striking. The bower is typically oriented north-south and if it is made to deviate experimentally, the bird will re-estab-lish the original orientation. Certain bowerbirds prefer blue objects; the dominant males, who are blue, will steal them from young subordinate males, who are green. A male bowerbird will choose blue cards and reject red ones; among different blues, it will choose the one of greatest intensity. Many investigators have remarked that the favor-ite shade of blue closely resembles the eye of the female, which is a brilliant blue. The female, it should be noted, takes no interest in the colored objects, but rather in the dances and raucous cries of the male.

Certain species decorate their nests, a practice that may be very elaborate in individual birds. Some apply the pulp of blue berries from various plants to the inside walls; others use bits of charcoal which they make into a paste with their saliva. Still others fashion a tampon or "paintbrush" from woody fibers they have shredded; holding this instrument in its beak, a bird will use it to smear the walls of its nest with the charcoal paste it has prepared. Females and some males never paint.

Other bowerbirds construct a veritable hut with a center post where they hang various objects, sometimes by wrapping a spider web around them. *Amblyornis subalaris* is reportedly capable of placing a center beam against a tree and making it stick fast by smearing it with a sort of resin. A completed hut has two posts which are wrapped with fibers up to a height of 3 meters; the bird inserts flowers, lichens, or colorful berries in the fibers. The flowers are usually white orchids which may continue to grow in the hut. Bowerbirds are very particular that the corolla be upright; if the investigator turns a flower downward, the bird becomes extremely agitated and turns it up again. These sometimes enormous huts seem to be constructed by several birds and perhaps there exists an embryonic society.

6. Conclusions

It certainly seems that in investigating construction we may be approaching the heights of psychic activity in animals. The study of animal building behavior is most fruitful, however, when one observes the following rules.

1. *The rule of the unusual.* The more unusual the situation, or the less likely it is to be met in nature, the more variable will be the animal's behavior and the more likely

that its adaptation will rise above the stereotyped behavior of the species. As Lecomte said, the titmouse *knows* how to build its nest, but it *seeks* the solution to a repair problem.

2. *The significance of the problem.* The animal must take an interest in the problem and therefore it must be significant to the animal. Problems that interest a spider are those expressed in terms of filaments; for a bee, they must be expressed in terms of wax, in terms of branches for a beaver, twigs for an ant, paper for a wasp, etc.

3. *The rule of individual variation.* We are dealing with the highest levels of psychic activity, which only some individuals can attain; hence statistical studies are meaningless.

4. *The rule of greater complexity.* It is the *most complex nest, not the simplest one*, that offers the most opportunities to present problems and elicit solutions. For the higher capabilities of the psychic equipment of animals

Figure 8. A beaver building its dam. (After Richard, 1964.)

to be observed, there must be an opportunity for those capabilities to be manifested; they can hardly be revealed if the nest consists of a simple hole in the ground or a haphazard pile of twigs.

CHAPTER 3

The Height of Animal Learning:
The Case of the Primates

It must first be pointed out that anthropoids are not necessarily capable of rapid learning. They may need as many as several hundred tries before they learn a single and seemingly simple example of visual discrimination. It seems that the further removed the response is from the stimulus, the more difficult the learning process becomes for the animal. For example, if pieces of bread are dyed different colors, some of which characterize inedibility (e.g., bread soaked with quinine), primates find the problem very easy and learn rapidly. If, on the other hand, undyed pieces of bread are placed in color-cued plastic cups, they find differentiation of the bread according to the color of the cup much more difficult.

1. The Higher Faculties of Apes and Monkeys

a. *Learning relationships.* A test often administered to simians and other animals involves learning a relationship. For example, a large box A should be preferred to a smaller box B; in the next step, however, box A is presented along

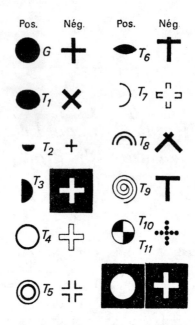

Figure 9. Transposition experiments with chickens. First the two symbols on the left are used for training, one being positive and the other negative, then the chicken must transpose its learning to the two symbols on the right. (After Schulte, 1970.)

with a still larger box C, and it is then C that should be chosen. This phenomenon of transposition was questioned by strict associationists, who considered it impossible or at least very rare. But it does indeed exist (Gonzales et al., 1967) and in a number of species.

    b. *Learning sets.* Harlow and Warren (1952) made great strides regarding the problem of insight as propounded by Köhler (1921)[1] by studying the effect of long learning sets in which the animal must follow the same rule

---

[1] Insight or sudden understanding of a problem was formerly considered proof of "intelligence."

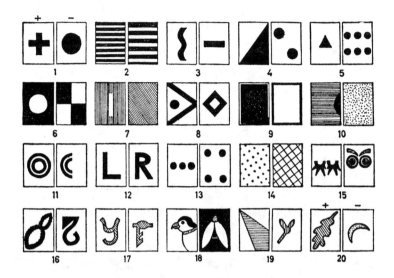

Figure 10. Capacity for visual discrimination in Indian elephant. In each set of symbols, one is positive and the other negative. (After Rensch and Altevogt, 1953.)

although the object to be chosen varies. A point is reached when choices will be successful at the first try and apes that use Thorndikian trial-and-error can be transformed into apes that use Köhler's insight. We should add that Köhler himself never denied the influence of learning or, more precisely, prior experience, on primate behavior, including insight, but perhaps he did not make that clear enough in his publications.

c. *Singularity*. In tests involving the choice of an object as being singular, that is, different from the other objects, one achieves remarkable success with macaques, orangutans, and chimpanzees, while one obtains poor results with cats. Chimpanzees and young children greatly surpass other primates in this test.

d. *Generalization*. In recent years much progress has been made with regard to the capacity for generalization. There had been a question whether monkeys or rats were capable not only of learning to recognize a triangle but, in addition, of perceiving "triangularity" when the triangle was presented in different positions and different surroundings. Rensch (1973) shows that in all vertebrates with a good capacity for visual learning there are similar tendencies toward generalization. These animals behave as if they had established a nonverbal generalization; a figure is recognized irrespective of whether it is enlarged, reduced, turned around, presented only through some of its components, or changed in color, as long as its basic characteristics are preserved. At times the capacity for generalization is carried very far. Rensch and his colleague Dücker (1959) succeeded in teaching a civet cat (*Viverricula*) to respond to all sorts of figures that varied widely in shape and background, according to *whether or not there were curved lines in the figure.* Thus it is easier to understand the experiment of Herrnstein and Loveland (1964) in which they taught pigeons to recognize *the presence or absence of human beings* in scenes in photographs; it might be said that the pigeons had formed the concept of person.

Rensch (1973) also shows that a great many animals, not only primates but also fish, are capable of distinguishing elementary quantities, for example, 2 objects from 3, or 3 from 4, and they can do so irrespective of transposition or permutation of the objects. Without going so far as the numerical abilities of Koehler's (1954) birds, it must be recognized that a primitive numerical capacity is, in fact, universally distributed. More broadly it may be called, after Rensch, a capacity to form nonverbal concepts. And there are other examples of this, such as the concepts of odd and even or equal and unequal, which are also readily absorbed by many animals.

This leads Rensch to wonder whether animals might not be able to acquire even more complex concepts, such as forming some perception—more or less vague—of "self" (*Ich-Begriff*). It should not be forgotten that even in children this concept develops only little by little. "Self" is understood as that which remains one and the same through changing events and, for men, "that which thinks." There is a primary ego (the verbal or nonverbal conception of the constant through the inconstant) and a secondary ego (the conception of thought as the thinking subject). Obviously, only the first is possible in animals.

As a child becomes conscious of the movements of his own body and aware of the universe as a separate object that he cannot move so easily, he comes to perceive himself, or rather feel himself, to be a distinct object. The process is probably very similar in apes. The fact that many mammals and birds can *learn to recognize their name* also points to some conception of self. An even stronger argument can be found in the behavior of Washoe and Sarah, chimpanzees who can name themselves. The mirror experiment, which has a fascination for all primates, is also worthy of note. Chimpanzees are able to utilize a mirror in extracting a bothersome food particle from the teeth. After having anesthetized a chimpanzee, Gallup (1970) put red paint on its head; upon awakening, the animal inspected the red spots on its head in the mirror. One finds no behavior of this sort in monkeys such as macaques. As for children, they do not being to react to a mirror until about 10 months of age.

e. *Observation*. Learning through observation seems to exist in chimpanzees or, more precisely, it can be elicited experimentally. Hayes and Hayes (1952) compared their

Figure 11. An orangutan looking in a mirror places a lettuce leaf on its head. In the opinion of Lethmate and Dücker (1973), this implies a certain degree of self-recognition or self-awareness (*Ich-Begriff*).

young child and Vicki, a female chimpanzee, in learning actions demonstrated by the experimenter. The chimpanzee proved to be at least as skillful as the child in this exercise; in one case it even succeeded in learning by observation where the child failed. Learning through observation also exists in rodents (Pallaud, 1972).

f. *Cooperation.* The question of cooperation was examined by Crawford in 1937. Several chimpanzees had to pull on a rope together in order to obtain a very heavy box containing food. It required some time and the help of the investigator before cooperative behavior began to emerge, but in the end cooperation became very marked. It was found that the subordinate animal goes and seeks the aid of the dominant ones;[2] there is, moreover, a veritable pantomime of solicitation. It is curious to note that if the cooperating animals are switched to another task, where it is no longer a matter of pulling on a rope but one of pushing on a lever that is hard to depress, then cooperation ceases; it is closely dependent on the context and environment.

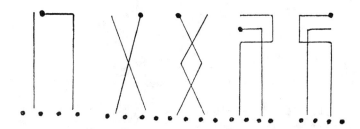

Figure 12. String problems. By pulling one of the strings the subject can obtain food, represented by a black dot. In order to succeed, an animal must clearly perceive the geometric continuity. Gorillas have much more difficulty solving the problem at the far right than do chimpanzees. (After Riesen et al., 1953.)

g. *Food tokens.* In their famous experiments, Wolfe (1936) and Cowles (1937) demonstrated that chimpanzees can learn to differentiate position, color, design, relative size, and duration, with *their sole reward being tokens*, which they can later exchange for food. They are capable of distinguishing tokens of different denominations, that is,

[2] It is the opposite in rats. In a study of rats that were raised in a terrarium and had to push a lever to obtain food, Mme. Anthouard (1969) shows that it is the subordinate ones that push the lever while the dominant one simply takes the food pellets that are dispensed away from them. If a subordinate does not want to work, the dominant rat goes to fetch him with a little gesture of solicitation; if the subordinate stills tries to shirk, he will be punished. Similarly, if the lever is at one end of a long cage and pressure on it dispenses food at the other end, it is always the same rats that push the lever though they receive no immediate reward because the others are busy eating at the other end.

Figure 13. A chimpanzee "fishing" for termites with a twig. (After a photo, van Lawick-Goodall, 1971.)

Figure 14. A chimpanzee using a stick to strike a stuffed leopard. (After Kortlandt, 1967.)

tokens that will yield more or less food when exchanged; they can also differentiate between food tokens and other tokens with no exchange value. They prefer the food itself, however, if it is offered at the same time as the tokens. They will also hoard food tokens and even steal them from one another.

h. *Retention*. Yerkes and Yerkes were the first to observe that when chimpanzees see a box baited with a tidbit and then closed, they have no trouble picking that box out from three others some 3 hours later. The limit of retention is about 48 hours. It is the position of the box that they memorize.

i. *Utilization of tools*. Chimpanzees use sticks for attack and defense and straws to "fish" for termites. Schiller (1949) contributed a great deal to interpreting simian behavior when he found that young animals will stack boxes on top of one another or fit the segments of a fishing pole together even without a reward. Klüver (1937) found that American *Cebus* monkeys also make excellent use of implements; however, primitive monkeys other than the *Cebus* do not use tools.

j. *Utilization of numbers*. The first experiments to find out whether chimpanzees can learn to use numbers (Douglas and Whitty) did not meet with great success. Nevertheless, Hicks showed that they can acquire the notion of "triplicity," and Rohles and Devine that they are capable of distinguishing the middle object in a row of several (up to 17). In their opinion, this indicates a certain numerical capacity, especially since the objects were not spaced regularly and there were gaps between them. It was Ferster and Hammer, however, who proved in an entirely automated experiment that chimpanzees can learn binary enumeration. The subjects could see three lights that were

Figure 15. The chimpanzee Sultan fitting together two sections of a fishing pole in order to get a tidbit that cannot be reached with only one section. (After Köhler, 1921.)

Figure 16. Left: a chimpanzee has stacked up three boxes to reach a banana. (After Köhler, 1921.) Center: four boxes stacked up by another chimpanzee in the same test. (After Bingham and Yerkes.) Right: stacked boxes combined with use of a stick. (After Köhler, 1921.)

Figure 17. Solution of a complex problem by a pied woodpecker. A tidbit was placed in a test tube too deep for the woodpecker to get at it. Top: the bird pecks at the bottom of the tube which results in raising it a little. Center: the bird pulls the tube up. Bottom: the bird turns the tube and by tipping it far enough succeeds in getting the tidbit. (After photos, B. Chauvin-Muckensturm, 1973.)

turned on or off in irregular patterns (three lights off = 000; two off and one on =001, etc.). The subjects had to reproduce the models by using three switches that controlled three other lights in front of them. After 170,000 tries, the two chimpanzees that had undergone training could "write" the binary number corresponding to a given number of objects placed before them.

One wonders, however, if primates could go as far as Koehler's (1954) crows. These birds could recognize a given number of spots (up to eight) placed on the lid of a container that would yield food when opened. They succeeded irrespective of the color, shape, or arrangement of the spots. By consulting a picture showing a certain number of dots, they were also capable of going to look for food in the fourth or sixth container in a row, for example.

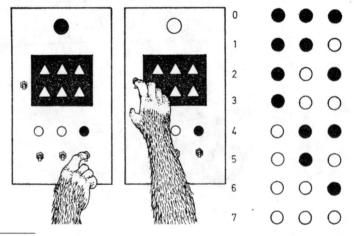

Figure 18. By looking at numbers in a binary system at the right (in the form of lights that are on or off), a chimpanzee first learns to reproduce them on a board in front of him by pressing the corresponding buttons; when he is later shown a certain number of triangles, he is able to indicate their number in the binary system on the board. (After Ferster, 1964.)

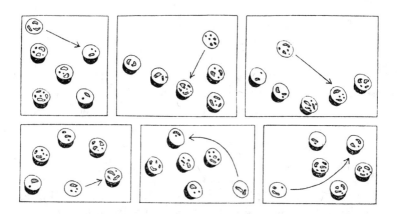

Figure 19. A series of correct choices made by a crow in accordance with a model (shown without shadow). Note that in terms of arrangement and size, the spots on the model are unrelated to the spots on lid of the jar the crow must open to obtain a tidbit. The number of spots is all that matters. (after Koehler, 1954.)

Figure 20. A jackdaw lifting the lid of a jar that bears the same number of spots as the model in the foreground. (After Koehler, 1954.)

2. SIMIAN LEARNING OF HUMAN
MODES OF COMMUNICATION

The research of the Gardners and Premack. A few
years ago, Gardner and Gardner (1969), an American
couple, taught a female chimpanzee, Washoe, to use
American Sign Language (A.S.L.), which is used by deaf-
mutes. At the same time, Premack (1970) was teaching
another chimpanzee, Sarah, to communicate in somewhat
the same way by using plastic symbols that she could stick to
a portable slate to form sentences. Before discussing the far-
reaching implications of these experiments in the realm of
animal language, we should like to note that such important results might have been anticipated in the light of what
was already known about the capabilities of animals. For
example, Herrnstein and Loveland (1964) had shown that
pigeons could respond to the presence or absence of human
beings in a photograph; they pecked on one disk if there
was a human being and on another disk if there was not.
Lehr (1967) carried out an analogous experiment with primates.

a. *The Gardners' work.* In view of the fact, well
known to ethologists, that the grunting vocal manifestations
of anthropoids express little more than emotional states and
it is impossible to transform them into a true vocal language, the Gardners proceeded to teach American Sign
Language to their chimpanzee. The acquisition of A.S.L.
was quite slow at first, then more rapid. The fact should be
stressed that on several occasions veritable "inventions" on
the part of Washoe were observed. At one point, for
example, the investigators were trying to teach Washoe the
word *apron*, which is expressed in A.S.L. by rubbing the
mouth with an open palm. An apron was held out to

Figure 21. Chimpanzees seem to enjoy drawing. Here Julia is busy scribbling. (After a photo, Rensch, 1965.)

Washoe to elicit from her the corresponding A.S.L. sign. She had forgotten it and tried several signs, then, in the end, using both hands she sketched the outline of an apron over her body, reaching behind her neck where the strings are tied and then down along each side of her chest!

The vital question is not whether Washoe has acquired the linguistic ability of an adult deaf-mute (certainly not!), but whether she can express herself as well as a child learning sign language. According to the Gardners, there is a close resemblance between the mechanisms available to chimpanzees and children.

b. *Premack's work.* Premack worked exclusively with metal-backed plastic symbols of different shapes and colors

Figure 22. When drawing, chimpanzees tend to follow the contours of a geometric figure that is given to them to start with. After Rensch, who wonders whether this might not imply the rudiments of an esthetic sense.

that could be affixed to a magnetized slate. Sarah was taught to arrange consecutive words one after the other vertically rather than horizontally, although it does not seem to make much difference to a chimpanzee. Premack believes that his technique is superior to others, for example that of the Gardners, because it is much easier to determine whether or not a symbol has been correctly placed on the board than it is to decide whether or not a certain gesture has been made. Furthermore, one can limit the number of symbols available to the chimpanzee at any one time, thus reducing the number of possible combinations and, perhaps, facilitating the task.

Sarah eventually succeeded in acquiring 127 "words," with which, 75-80 % of the time, she can complete or correct imperative, declarative, or interrogative sentences. In more complex cases, she succeeds in placing a string of words in the correct order. Neither Sarah nor Washoe can be induced to work unless they are on very friendly terms with the investigator.

The imperative mode. To begin with, Sarah was taught words such as *apple, banana, give, take,* and her own name *Sarah.* At first the investigator would give her a piece of fruit she liked. After a while, when she had begun to imagine that the distribution of fruit would be unending, she had to *request* the fruit in order to get it. For example, she had to use the sentence *Mary give apple Sarah.* In this simple sentence, whose acquisition was the learning goal, it is possible to change the identity of the donor, that of the recipient, and the name of the object given (for example, *Randy give apple Mary*).

Sarah was taught only one word at a time. She would first be shown an apple and at the same time the symbol for apple (a blue triangle). She would not be given the apple until she had placed the corresponding symbol correctly. Her learning was checked by offering her a piece of fruit and at the same time several symbols from which she had to choose the right one. Similarly, after a time, she was required to choose from among several symbols the one that corresponded to the fruit desired, which only then would be given to her. Later, the investigator would be changed and she would be taught to designate him with the corresponding symbol. In the end Sarah was able to formulate the sentence correctly.

Word order. This is obviously more complicated. Premack set out to teach Sarah the sentence *Sarah insert banana pail apple dish.* (She was to place the corresponding symbols on the board.)

She was first taught:

> Sarah insert banana pail.
> Sarah insert apple pail
> Sarah insert banana dish.
> Sarah insert apple dish.

She learned to form these sentences separately and they were then presented to her in pairs in all possible combinations, so that she might become used to handling two objects in one experiment. Next she was given the task of performing these two operations, but one after another in a specified order, for example:

> Sarah insert banana pail Sarah insert apple dish.
> Sarah insert banana pail insert apple dish.
> Sarah insert banana pail apple dish.

The fact that first the symbol *Sarah* and then the symbol *insert* were removed did not bother the chimpanzee.

The interrogative mode. Premack began by teaching Sarah the terms *same* and *different,* which provide very economically the makings for interrogation. Two similar objects (two cups or two spoons) were put before her and she was given a plastic symbol—*same*—which she was to place between these two objects. In the next phase of the experiment, she was shown a cup and a spoon and given the symbol *different.* Finally, she was shown two cups or two spoons and given both symbols at once: she had to choose. After five tries, she succeeded 80% of the time. The experiment was then repeated with objects she was not used to: a peg and a ball, for example. After 20 "transfers" of this type, she was successful 80% of the time.

Premack went a step further by introducing the *question mark*. It is a special symbol that was placed between two objects presented to Sarah. She had to remove it and put in its place either the symbol *same* or the symbol *different*.

Yes or no. To teach *yes* and *no*, the question mark was placed after two objects and between them was the sign *same* or the sign *different*. The chimpanzee then had to affix one of the two symbols, *yes* or *no*, according to whether or not she agreed with the sentence written on the board. This obviously implies that the animal has managed to abstract the question mark to some extent and generalize it. Without that, the experiment would be impossible. It indeed seems that Sarah generalizes and that she is capable of abstraction. Later on she also had to generalize the symbols *yes* and *no*.

Learning names. Sarah was given a real apple and the symbol *apple* (a blue triangle!) and between the two was placed the symbol for a question. Sarah had to remove it and insert the symbol *name of*. This was repeated with a banana and the special symbol representing it. Then Sarah was presented with a banana along with the word *apple* and the *question mark* symbol in between. She had to remove that symbol and replace it by the negative sign coupled with the word *name of*, i.e., *not the name of*. After this, the investigator, as usual, went on to transfers with other fruits and other objects. The symbol *name of* has now been completely generalized, and when Premack wants to teach Sarah a new word he shows her the object, its symbol, and the word *name of* in between. Then he presents the object with a symbol that does not match it and inserts *not name of* between the two.

Result. Included in Sarah's vocabulary of more than

120 words are nouns, expressions such as *color of* and *shape of*, five verbs (*cut, eat, give, put, take*), adjectives (especially of color and shape), prepositions (*in, on, in front of, to the side of*), conjunctions (*if, then*), etc.

    c. *The "Klüge Hans" hypothesis.* This hypothesis is named after Clever Hans, one of the Elberfeld horses which were supposedly capable of extraordinary feats, such as extracting square roots! Actually, in most cases it was probably a matter of signals unconsciously transmitted by the trainer to the subject (for example, a change in breathing, tapping the foot, etc.). In the case of Sarah, and especially Washoe, cannot the question be raised of the existence of similar phenomena? Might not the investigator unwittingly indicate to the chimpanzee what it should do?

    The double blind. The best test is that of the double blind, in which one assistant announces the results of the test to another who records them, without either one of them being aware of what is really being sought. Premack did not carry out this test exactly, but he did use the following techniques.

    In a first attempt, after placing 10 questions on the board, the investigator left the room and Sarah was supposed to answer the questions with *yes* or *no*. That failed—she refused to work in the absence of the usual investigator. Premack then introduced a new examiner who made friends with Sarah but did not know the language. He was given a series of numbers, each of which corresponded to a symbol, and all his instructions about how he should place the symbols were in numerical form. He used a portable phone to report the numbers of the symbols placed by Sarah to another assistant in a different room. The latter translated the figures and Sarah was then rewarded or not, depending on whether or not her response had been satisfactory.

Under these conditions, Sarah performed relatively poorly. Instead of an 80% success rate, she scored only 70%, which was still well above the level of chance (13-20%). She showed a sort of regression to an earlier stage, placing the right words, for example, but at random instead of in the proper order. It would seem that emotional factors were to blame for the increase of errors. Premack wonders, though, how a human subject would react in such circumstances, i.e., if he were asked to "talk" to a deaf person who would then supposedly transmit the information, at least in part, to a third person who understands the language. In addition, the fact remains that, as was bound to happen, the blind examiner invariably formed hypotheses about the meaning of the experiment, understanding some aspects of it correctly and others incorrectly. Therefore the precaution was taken of scoring only the first part of each session, when the examiner had not yet had time to deduce very much.

The problem in the case of Washoe. The Gardners, who taught their chimpanzee, Washoe, American Sign Language, also attempted verification by the double-blind method. Washoe had to describe photographs to an examiner who could not see them, and he in turn interpreted what she had conveyed to him. In addition, other tests were carried out using one-way glass so the examiner could see Washoe but she could not see him. Also, two deaf observers were able to understand Washoe the first time they saw her with 70% accuracy, and 95% the second time.

The Clever Hans objection has been definitively disproved, however, by Rumbaugh and his colleagues (1973), who carried out a Premack-type training program with another female chimpanzee using a computer. She must press keys bearing the appropriate symbols and do so in the right order; only then is she rewarded. All the results are recorded and an examiner is never present; nevertheless, the results are very good.

Semantics in Washoe and Sarah. According to Premack, Sarah shares at least one capacity with man, that of breaking down a global visual experience into separate components. On the other hand, Washoe showed the ability to apply a concept to a still larger class of related elements: she used *flower* to designate not only a flower but also a tobacco pouch, a mentholated salve, and other odiferous things. Only later and with help did she dissociate *smell* and *flower* and apply the latter correctly.

As for what conception Sarah may have of words, it should be remarked that she is not at all confused when she is asked to put the *name of* some object rather than the object itself in a pail.

Analysis of properties. One of Sarah's most astonishing abilities is no doubt that of enumerating the properties of an object. For example, she may be presented with an apple and two alternatives: the symbol for *red* (a red disc) or the symbol for *green* (a green disc). She will place the red one on the board. Between the symbol *round* and the symbol *square*, she chooses the round one and similarly attributes a stem to an apple. If instead of a red apple she is presented with the symbol for *apple*, which is, it should be remembered, a blue triangle, *she responds in the same way; thus she does not analyze its form but what the form represents.* She was given the same problem with real objects and symbols for *candy* and *Mary*, and was equally successful in isolating their properties when presented with either the object or the symbol that represents that object.

It is curious to note that Sarah failed to recognize photos of apples in magazines as apples, though it is true the photos varied widely in format and context; also, Sarah's training did not include a symbol for *picture of* or *reproduction of*. On the other hand, Washoe readily recognized photos of cats, as did Kiki, the chimpanzee of Hayes and Hayes (1952). But Kiki and Washoe, unlike Sarah, had had picture books available from an early age.

Quantifiers. Sarah has learned the meaning of the words *all, several, one*, and *none*. She can understand and form sentences such as *some circles are green* and *one cookie is big*.

Syntactical ability. Obviously, the crucial question was whether Sarah could write sentences *in the proper order.* She learned to do so, but with some difficulty; in the beginning her sentences were poorly constructed. Some have maintained that Washoe pays no attention to word order. The Gardners, however, claim that she is at least as careful about syntax as children are at a comparable stage of development.

As for *interrogation*, Premack has not yet carried his experimentation to the point where his chimpanzee is given the opportunity to ask questions herself.

The verb "to be." Starting from the sentence *Mary give banana Sarah*, Premack removed the symbol *banana* and introduced in its place a new symbol for *fruit*; then when a variety of fruits were shown to Sarah—apple, banana, and orange—she had to write *Mary give fruit Sarah* before she was rewarded. This in turn permitted another new symbol, *is*, to be introduced: Sarah was taught to write *banana is fruit*. The verb "to be" was then transferred to other situations, for example, *red is color* or even *triangle is not color*. Later a symbol for *plural* was introduced and

Sarah learned to write such sentences as *banana orange apple are (is + plural) fruit*.

The negative. With respect to the negative, Sarah's behavior was very curious. The examiner might place a red card on a yellow card, for instance, and then ask *red on green question mark* ("Is the red on the green?"). Sarah would correctly reply *no* 65% of the time. The rest of the time, however, she would remove the yellow card, substitute a green one, and then answer *yes!* This is very difficult to interpret.

In any case, Sarah prefers affirmation to negation. Also, she is reluctant to write sentences describing situations she does not like. To get her to write *Mary give apple Randy* —which meant she would not get the apple herself—she had to be given a tidbit she liked better.

Possible developments. Eventually Premack would like to bring Sarah to understand and use the terms *why* and *because* in sentences such as *Mary not give apple Sarah because Sarah naughty*. He would also like to introduce a kind of "morality," that is, to inculcate in Sarah an inventory of actions labeled *good* or *bad*. Another goal is to teach her the comparative, such as *bigger than*, along with arithmetic, or at least the terms *more*, *less*, and *equal to*. All that should not be impossible for, as we have seen, it has been proved that chimpanzees can count in the binary system.

d. *Can one speak of simian language?* First of all, it is hard to agree on a definition of language. Ploog and Melnechuk (1971) summarize the various positions as follows:

1. Some contend that all overt behavior is communicative in nature; others maintain that communicative behavior can be distinguished from noncommunicative

behavior. (There is at least as much ambiguity about the term "communication" as about the term "language.")

2. Some stress the vocal nature of language, others its capacity to designate and to formulate affirmative, interrogative, and other propositions.

3. There are those who emphasize the utility of language for communication while others focus on its abstract nature. According to Lenneberg (1964), just as arithmetic does not consist solely in solving equations, language cannot be basically defined as an exchange of information; it constitutes a particular form of relations.

4. Some would reduce language to a set of Skinnerian-type reactions: differentiated stimulus ⟶ response ⟶ reinforcement. Others, for instance linguists, analyze it in terms of rules the speaker must know in order to speak it, but they are subject to exceptions (codified) and cannot be considered fundamental principles.

5. Regarding these rules, some lay stress on the progress made in recent years in cataloguing rules and establishing their hierarchy; others, however, doubt that such rules even exist. Chomsky's (1965) *generative grammar* is a system intended to "generate" all the sentences that speakers find "well formed" in their particular language. But those who are trying to devise programs for translating machines say that what they need to know is how a language is organized to convey meaning, rather than how isolated syntactical structures are organized.

6. Some are interested more in the generation of language, others in its interpretation. The latter note that the first thing a child must do is understand what is said to him, even before he speaks; hence interpretation comes well before generation of sentences.

7. There are some who make a distinction between

semantics and syntax and others who find they cannot be separated. Semantics is concerned with the meaning of words and sentences and syntax with the arrangement of words within sentences. It is true that sometimes meaning depends on the word and other times on the word order; a strict dichotomy between semantics and syntax would seem impracticable.

8. Some emphasize necessary characteristics that are inherent in any language. Others note that none of these characteristics have any meaning when considered alone and the fact that one or another of them may be lacking does not necessarily mean that we are not dealing with a language. As Lenneberg says, the use of criteria is perhaps rash, for language is not simply the sum of its components, it is a "system."

What conclusions can be reached? It is no doubt premature to pose such a question. Hebb and Thompson (1954) suggest a dual definition of language: first, it consists of combining two (or more) gestures or sounds to produce a single effect; second, it uses the same gesture in different combinations in order to obtain different effects, readily changing them to suit the circumstances. The types of communication used not only by Washoe and Sarah but also by a number of birds clearly fall within those categories. It is true that there remains the question of intention, whether the animal acts "with a view" to something. That is a rather specious question, however, for anyone who has truly worked with primates, dogs, or other higher animals.[3]

---

[3] Chauvin-Muckensturm (1973) has revealed in the pied woodpecker (*Dendrocopus*) the existence of phenomena utilizing a sound code, rather close, no doubt, to what occurs in Washoe and Sarah.

CHAPTER 4

Migration and Orientation

1. Bird Migration

Migratory birds sometimes cover enormous distances. The champion would seem to be the Arctic tern (*Sterna paradisaea*) which leaves the Arctic seas in late summer and travels some 18,000 kms. to the coasts of Antarctica to spend the winter; then in the spring, it returns to its nesting grounds in the Arctic. In effect, then, it completely circles the globe, whereas a good carrier pigeon rarely exceeds 1,500 kms. in its flight home. Between those two, one can cite an albatross that was released 6,500 kms. from home, and returned to its nest in 13 days (Kenyon and Rice, 1958). Such phenomena raise a number of questions which researchers have not fully answered.

a. *Can birds recognize the regions they overfly?* There are arguments for the affirmative, notably the remarkable visual memory of the homing pigeon. Skinner (1951) has shown that this bird can be taught to select a precise point on an aerial photograph and it will still remember it four

years later.[1] Also, learning markedly improves the performance of homing pigeons. Further evidence is provided
by Griffin and Hock (1949) who released some gannets 340
kms. from their nesting area. Following the birds in an
airplane, they found that their route was very sinuous,
as if the birds were looking for reference points on the
ground. It should be pointed out, however, that other
experiments, while not rejecting the possibility or utility of
terrestrial reference points, show that they certainly do not
play an exclusive role in migration. For example, take the
experiment of Perdeck (1958). He moved some starlings
from Holland to Switzerland and released them at the time
of their fall migration. The birds started south as usual, but
the adults corrected their route by veering westward, while
the young birds followed a path parallel to the normal fly-
way from Holland (see Fig. 23). This clearly shows that
they must possess a program that tells them the general line
to follow but not the precise itinerary. In many species,
though, the year's young make the first migration of
their lives either before or after the adults. In numerous
cases, therefore, it is obviously not necessary to learn the
trajectory. Furthermore, "migration corridors," which
were formerly thought to be quite narrow, can actually be
very broad, up to hundreds of kilometers wide at times.

  b. *Do birds use the sun for orientation?* This theory
goes back to Kramer (1949), who observed that the better
the visibility of the sun, the better the orientation of birds.

---

  [1] This astonishing capacity of the homing pigeon led Skinner (1953) to en-
visage a *pigeon bombardier*. The bird would be installed in an airplane that
would be guided by remote control to fly over a point where a bomb was to be
released. The pigeon would have been trained to peck on the image of the target
when it appeared in the center of a screen showing a televised picture of the
ground—and that would trigger a bomb release!

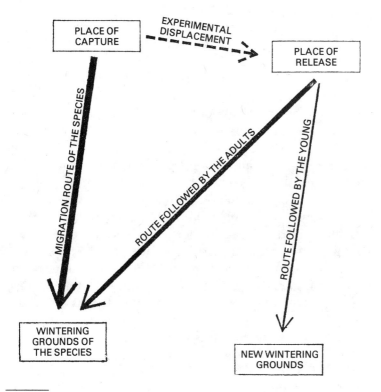

Figure 23. Displacement experiment of Perdeck (1958). Migratory birds were captured at their point of departure, then released southeast of the place of capture. The adult birds flew to their usual wintering grounds in the southwest. The young birds, released at the same time as the adults, followed a trajectory parallel to the normal route which took them the same distance in the same direction, south-southwest, but not to the same place.

By a play of mirrors reflecting the sun, he even succeeded in deflecting the orientation of birds with respect to the ground. Matthews (1955) later carried this theory much farther. According to him: (1) a pigeon takes a reading of the height of the sun; (2) from that, it extrapolates the altitude of the sun at noon; (3) it remembers the altitude of the noonday sun over its nest, which is its destination; (4) thus it can determine its latitudinal displacement, e.g., if the deduced altitude is higher than the remembered altitude, then the bird is south of its nest; and (5) it can determine its longitudinal displacement by means of a precise biological clock pigeons are thought to possess (which can, in fact, be proved). Matthews maintains that the bird estimates the difference between the observed position of the sun and its projected noon position, on the one hand, and, on the other hand, the same measurements as remembered at the nesting site at the same time of day. In this way the pigeon would obtain an estimate of its longitude. Birds are also reported to be able to estimate altitude above sea level with good precision.[2]

It is certainly hard to believe that animals can accomplish such complicated calculations so automatically, and, as numerous critics have pointed out, it is especially hard to accept that they do so in some 30 seconds. Experiments of several investigators show that pigeons do indeed keep track of the sun, but in a far cruder fashion. Also, in view of the fact that many pigeon races take place at night, it is not necessary to dwell too long on Matthews's theory.

Be that as it may, *stars* can serve as reference points, as Sauer (1961) showed with *Sylvia* and Emlen (1969) with *Passerina cyanea*. These two species alter their orientation in a planetarium when the position of the stars is changed and, as long as they can see the stars, they tend to be unaffected by changes in the magnetic field. It is therefore

[2] Lehner and Dennis have shown that mallards can detect minute changes of pressure on the order of 0.4 lb. per square inch (conditioned reflex technique); thus the bird could detect the approach of a storm from 6 to 10 hours in advance and register a change of altitude of 1,000 feet.

the panorama of the night sky, no doubt, that enables them to migrate over the earth from north to south without becoming too disoriented with respect to the directional change of the magnetic field.

c. *The influence of meteorological factors.* No one doubts the influence of the weather, at least in triggering migration. Spring migration usually takes place after the arrival of warm fronts associated with a high pressure formation; it is the opposite for fall migration. When an unusual meteorological condition occurs (a warm front in the fall, for example), birds may well reverse their route and fly in the opposite direction to normal for a few days. But Nisbet and Drury (1967) have revealed a still stranger phenomenon: the number of birds that depart is closely related to *the weather at the point of destination!* If the weather (at the place of arrival) is dry and rainless, with a moderate temperature and high clouds, the rate of departure will be very high. There is no explanation for this astonishing phenomenon.

d. *The magnetic field and bird orientation.* Along with an increasing number of writers, Southern (1970) points out that the question of magnetic orientation in birds must be reconsidered. In gulls (*Larus delawarensis*), he noted a basic preference for the direction of their seasonal migration, that is, toward the southeast in the area in which he made his study. During magnetic storms, however, their ability to orient themselves in cages disappeared and the birds turned in random directions. Stutz, on the other hand, discovered in gerbils (*Meriones unguiculatus*) regular fluctuations in spontaneous activity that could only be correlated with regular changes in the horizontal magnetic component at 11 a.m., at 2 p.m., and from 3 to 6 p.m.

According to Keeton, the difficulties encountered in measuring the importance of magnetic factors in the orientation of pigeons arise from the fact that *in clear or sunny weather magnets attached directly to the birds have no effect on them, whereas in completely overcast weather birds bearing magnets often become disoriented*; control birds, however, bearing metal bars of the same weight, return to the loft without difficulty! But Keeton has shown something more: in the case of pigeons never having left the loft before, magnets often destroy all capacity for orientation, even if there is sun; young birds therefore utilize orientation in a different way than do adult birds.

According to Wiltschko (1968), birds turn as the magnetic field turns and become disoriented when its strength falls by a quarter or rises by half; if they are exposed to truly perpendicular lines of force so that the angles between the magnetic field and gravity are equal in the north and south, they leave for the north and south in equal numbers. This obviously raises questions concerning the behavior of migratory birds that cross the equator and must contend with a reversal of orientation.

Another very special complication arises from the fact that in certain very localized geographic areas, which are often well known to breeders, pigeons are incapable of orienting themselves by day. Yet these zones show no unusual geological or magnetic characteristics. Pigeons flying over these areas have been followed by plane; it was found that they fly completely at random.

On the other hand, birds crossing these zones at night or even *in heavily overcast weather* can maintain their sense of direction very well. The only clue we have concerning these zones is somewhat anecdotal. Keeton reports that in one of these areas, which is not far from Cornell, forest rangers experience great difficulty in maintaining radio communication between their vehicles; at times communication becomes completely impossible.

In conclusion it can be said that, far from using a single stimulus to return to the nest, birds undoubtedly use a great many sources of information, relying more on one or another depending on the circumstances.

2. ORIENTATION BY THE SUN

This subject has been particularly well researched in bees (see p. 196). Not only bees, however, but many other animals as well use the sun as their compass. Little crustaceans, such as *Talitrus, Talorschestia,* and *Orchestia,* use it to find their way back to the water from the land. For example, Pardi (1960) collected specimens of sand fleas from three Italian beaches having quite different orientations. He found that when these animals were transported to the laboratory and released under artificial light, they went in the direction of the native shoreline of the population to which they belonged. And their descendants do the same! In addition, if a sand flea from zone A is transported to zone B where the sea lies in a quite different direction, it will head in the direction of its native shoreline and hence will not find the water; it can be concluded that sand fleas have no perception, either direct or indirect, of the sea.

Orientation by the sun is also found in fish. In the experiments of Hasler (1960) and his colleagues, a fish was placed in a small plastic vessel in the middle of a circular enclosure and then induced by electric shock to take refuge in one of the several dark compartments surrounding its central position, which was open to the sky. The subjects used were fish from Lake Mendota (Wisconsin), which head for their spawning grounds near the shore when released in open water, except in overcast weather. Their orientation was as perfect in the laboratory as it is in nature.

The territory of turtles is barely 100 meters in breadth, yet if they are moved 5 miles away, most turtles will orient themselves in a homeward direction unless the sun is hidden or they are exposed to its reflection in a mirror (in which case they deviate by 180°). This is not, however, a matter of true orientation to the sun and the sun alone, for if they are moved 150 miles away, they become totally disoriented. This brings to mind the orientation phenomena observed in field mice by Bovet (1960) among others: in a circular maze they will find the exit that opens in the direction of their nest, except when it is more than 3 kms. away, which disorients them.

Orientation by the sun is certainly used by amphibians, but study of the phenomenon is complicated by the fact that extraocular orientation may supplement sight (Adler, 1963). Smell unquestionably plays a role, for blind *or* anosmic toads can orient themselves, but they become incapable of doing so if they are both blind *and* anosmic. No doubt their sense of smell is what enables amphibians to find their way home on rainy and moonless nights.

3. ORIENTATION BY GRAVITY

This type of orientation is very common. The coleopter *Bledius* finds the vertical with an accuracy of 1° to 2°. *Dyschirius*, another beetle, is capable of climbing upward on a plane inclined only 3° from the horizontal, and ants do almost the same (Markl, 1962). Honeybees manifest a very precise sense of gravity. When their comb is gradually inclined toward the horizontal, their tail-wagging dances are transformed into dances oriented directly toward the food source, without transposition of the angle; but for some dancers an inclination of only 5° from the horizontal is enough to elicit the dance with transposition. In insects the organs sensitive to gravity generally consist of hairy regions found in various parts of the body which alert the nervous system to changes of position of the various segments (Lindauer and Nedel, 1959; Schöne, 1959; Brückman, 1962).

4. Orientation by the Magnetic Field[3]

The earth's magnetic field is differentiated by intensity, which is specific to a given place and varies periodically with the time of day and season of year, and also by direction, with the lines of force of the terrestrial field spreading out from the magnetic north pole and coming together again at the south pole. These lines of force include a vertical component and a horizontal one, which vary with the geographical location.

Numerous observations have been made on the orientation of animals according to the magnetic field. Such orientation has been noted in paramecia and in *Volvox globator* (Kogan et al., 1966; Palmer). Many insects also use the magnetic field and, if this was late to be recognized, the reason is quite simple, namely, its *diurnal variations*, which were not taken into account. Adding up all the results obtained at no matter what time of day was bound to yield nothing. By carefully noting the hour of his observations, however, Becker discovered that cockchafers and wingless

---

[3] For birds see p. 81

imagos of various termites orient themselves according to the geomagnetic field; the same is true of the imagos of *Sarcophaga, Calliphora, Lucilia, Musca domestica*, etc., which in fair weather always land on plane surfaces in an east-west or north-south direction. The direction in which they alight remains the same irrespective of the position of the sun. Becker (1971) further found that termites of the genus *Reticulitermes* orient their galleries primarily in an east-west or north-south direction and this direction can be modified by introduction of a magnet. Finally, just recently, Oehmke (see Lindauer, 1971) has shown that if bees are enclosed in an opaque cylinder, they will build combs oriented in the direction of those in the hive they come from, but if the magnetic field is rotated 40°, the newly constructed combs in such a cylinder will correspond to this rotation. If the lines of force are distorted by placing a metal armature asymmetrically on one side of the cylinder only, combs will be built on one side only. In addition, von Frisch and Lindauer (1954) have shown that the small errors always observed in bee dances, especially at certain times of day, can be counteracted and changed by an artificial magnetic field.

Schneider (1961, 1963) found that cockchafers can orient themselves by the magnetic field alone when they are put in a dark cylinder, turning toward their feeding or egg-laying sites.

The Schneider effect. There are, however, irregularities due to an unknown field which traverses glass, walls, and Faraday cages. Schneider calls it an "ultraoptical" field, a rather unfortunate term since the phenomenon has nothing to do with light. Its disturbances are manifested as follows: although the magnetic field remains fixed, the cockchafer's orientation does not, at least not completely, for it will show preferences for certain sectors that are not always the same, but rotate with time in a clockwise (or sometimes counterclockwise) direction according to the *phases of the moon.*

Schneider therefore wondered whether the cockchafer might not be sensitive to *gravity*, which has variable local, solar, and lunar components. He found that a cockchafer modifies its orientation when a 40 kg. lump of lead, hidden behind a screen, is placed near the Petri dish where the experiment is taking place. The cockchafer "points" its head toward the lead (but not toward the screen if there is no lead behind it). Inasmuch as Schneider's research has been confirmed by other investigators, one may well ask whether disturbances connected with the phases of the moon—also noted by Birukow (1964) and Brown et al. (1964)—might not be related to a sense of gravity.

5. "Active" Orientation

In this type of orientation the animal actively emits various kinds of signals from which it or its fellows take their bearings.

a. *Electrical emissions.* Electrical eels and Mormyridae emit electrical impulses in the water and special receptors inform them of differences of resistance and the presence of obstacles in the vicinity (Lissmann, 1958; Lissmann and Machin, 1958).

b. *Odiferous emissions.* Under this rubric comes the odorous staking out of territory so frequently found among mammals (see p. 177). Such marking is also found in insects. *Bombus lapidarius* marks bushes at a height of 1 to 2 meters with secretions of a mandibular gland, while *Bombus lucoram* carries out the same operation at the tops of tall trees (Krüger, 1951). Females assemble on these scent trails, attracted by the odors deposited there by the males. In honeybees, the drones come together at one place, flying at a height of a few dozen meters; these "bee balls," well known to apiculturists, may be held in the same spot for several decades. Queens go there to mate, but it is not known whether the males mark the site with an odiferous substance. Lastly, in *Melipona*, a tropical American bee without stinger, honey gatherers have no dance to indicate the direction of a food source to their fellow foragers; rather, they deposit an odorous "Ariadne's thread" from their mandibular glands to mark the trail from the hive to the food source.

c. *Sound emissions.* One of the best known methods of orientation operates on the basis of sound emissions. By recording the activity waves of the tympanal nerve of noctuids, Roeder and his colleagues (1964) demonstrated that these moths can perceive bat calls at a distance of 30 meters or more; when flying in the open they can be seen to zigzag abruptly if the taped cry of a bat is broadcast. Emissions used in orientation are primarily ultrasonic. Certain tuids, Roeder (1964) and his colleagues demonstrated that scrambles reception in bats.

Griffin (1953) discovered that bats use ultrasonic emission like a veritable sonar system. There is debate among researchers about how this sonar works. Some believe it involves frequency modulation, others intensity modulation, and still others partial overlapping of signal and echo which would produce a pulsating effect that could be used in analyzing distance. In any event, echoes are localized through ear movements. Ultrasonic sounding is also employed by swallows to spot obstacles (Novick, 1959).

Dolphins are capable of a rich range of ultrasonic

emissions, up to 100,000 Hz. If a metal ring or a dead fish is quietly slipped into a dolphin's pool, impulses will increase to as many as several hundred per second until the dolphin comes in contact with the obstacle or fish. These emissions probably also serve for intercommunication among dolphins. If two dolphins in adjacent pools must press on a certain lever from a group of four to obtain a reward, they soon succeed in putting on an amazing performance. A light signal can be used to inform one dolphin which lever it should press; but the other dolphin facing its four levers has no such indicator light. The experiment is designed so that the reward can be obtained only if *both dolphins press the right lever at the same time*. This is patently impossible unless there is some sort of communication between the dolphins. Communication does indeed occur, for the experiment succeeds, *except when a soundproof screen is placed between the two pools*. The mode of communication is therefore acoustic, and dolphins must have specific signals to indicate directions such as up or down, right or left.

As for mammals other than dolphins, recent research on whales (Payne and McVay, 1971) shows that the sounds they emit at about 20 Hz are highly suited for traveling in a liquid medium. Theoretically, they can be heard for thousands of nautical miles, which would mean that all the whales in the same ocean form part of the same herd. Little background noise interferes with this signal, especially since the level of 20 Hz is below the noise produced by storms. Finally, at this level very little energy is lost by reflection on the bottom; dissipation with distance is also unusually low, 3 decibels in 5,600 miles. Thus it is a remarkably well-adapted signal.

d. *Passive orientation by chemical means.* By this we mean the automatic emission of odors without "intentional" deposit as discussed above. Odors of the opposite sex are often perceived very selectively and with fantastic sensitivity. According to Kalmus (1955), dogs can be trained to recognize and differentiate the odors of monozygotic twins. Mell found that hawk moths or Sphingidae can perceive females at distances as great as 11 kms. Miles and Beck believe that instead of a strictly olfactory organ, these moths possess a sort of "spectrometer" that detects vibrations of the odiferous molecules in the infrared range. It should be pointed out that at a radius of 11 kms. the concentration of odiferous matter is scarcely more than one molecule per cubic meter! This could not possibly indicate direction, that is, not unless there is recourse to some mechanism other than the usual sense of smell.

In addition, insects are attracted by olfactory stimuli to the animal hosts or plants on which they feed. Such examples are too numerous, however, to be mentioned here.

6. VARIATION AND RHYTHM IN ORIENTATION

Animals display not only activity rhythms but also orientation rhythms, such as Geisler (1961) observed in *Geotrupes sylvaticus*. This dung beetle manifests two activity peaks, one in the morning and one in the afternoon. In polarized light, moreover, the insect's path will form different angles with the plane of the light vibrations depending on the time of day, ranging from 0° to 90° to the right of the plane in the morning and from 0° to 90° to the left in the afternoon, which corresponds to an easterly direction in the morning and a westerly one in the afternoon. This phenomenon is clear-cut, however, only under the polarized light of a blue sky and not under a polarized screen.

Birukow (1964) found that *Calandra granaria*, a little grain-eating beetle, has a tendency to climb in a vertical plane (geonegativism), but this tendency varies with the time of day, being strongest at 4 p.m. The beetle's tendency to avoid light also varies, but it follows *the phases of the moon*: it is at a minimum at 6 a.m. during the full moon; at 4 a.m. before the last phase; at 10 p.m. after the last phase; at 12 noon during the new moon; at 8 p.m. before the first phase; at 2 a.m. after the first phase; and again at 6 a.m. with the full moon.

Brown et al. (1964) studied *Nassarius*, a gastropod, and *Dugesia*, a planarian. He found that in emerging from a vertical tube, *Nassarius* tends to bear toward the right in the morning, but toward the left during the rest of the day. Inclination to the left is at a maximum around noon. These tendencies can be countered magnetically; in addition, these snails seem to be able to perceive geomagnetic strength.

Brown also tested the effect of very low electrostatic fields. Electrostatically charged plates were moved at random, the observers not being informed of the changes (as was also the case in the magnetic field experiments). The planarians were found to react according to the placement of the plates: if the plates are to the south, the worms turn to the right, but to the left if the plates are to the north. This is true only in the morning, however, the directions being reversed in the afternoon. These worms are also capable of detecting very weak emissions of gamma radiation, and they move away from the source; at still lower strengths, they are reported to have a tendency to approach the source.

One may be surprised to find a lunar factor in biology, a phenomenon that would seem to be totally mythical, though peasants have always believed in it. But Birukow is not alone in having pointed it out; in the United States, Brown and his colleagues have also found it in mollusks and worms. They have even succeeded in discounting the influence of the tides or moonlight; rather, it is the moon as a heavenly body that exerts some action, no doubt through gravitational disturbances.

7. DIRECTIONAL PRECISION AND THE HOMING PHENOMENON

Rigorous precision is not always necessary in orientation. Saila and Shappy (1963) carried out computer simulations with respect to salmon and pigeons: they show that in assuming an almost random orientation, with a "bonus" of about 3% for each instance of pointing in the right direction, it is possible to obtain a return home, and in a rather short time at that.

Biotelemetric monitoring of the return of fish to their spawning grounds shows that the route is not always direct or precise; there are great individual variations. In the case of salmon, however, it was found that between the time a transmitter was implanted in a fish and the time it was caught again near the coast, the fish had swum in practically a straight line at an average speed of 23 to 60 kms. per day.

PART II

THE PAIR AND THE FAMILY

CHAPTER 5

Sexual Attraction and Mating Behavior

1. COURTSHIP DISPLAY

In higher animals such as birds, reproduction is often preceded by migration to traditional nesting grounds. The pituitary gland, activated by the increasing length of the day, is what triggers this return migration and in turn reactivates the genital glands by the secretion of sex hormones. Then begin the courtship displays in which the male "pays court" to the female, which can markedly accelerate maturation of her gonads if by chance she is not at the same stage as the male. This can be proved by various experiments, for example, by keeping male and female pigeons or other birds isolated in separate cages. When females are placed close to cages containing males but still not permitted to mate, these females become ready for reproduction sooner than females kept in strict isolation. Depending on the species, a more or less elaborate sexual display usually precedes copulation.

a. *The stickleback*. Before the mating season these fish live in groups, but as soon as it begins male sticklebacks be-

95

come solitary and stake out a territory. At this time their
eyes become a brilliant blue, their backs turn from brown-
ish to greenish, and their abdomens become red. Whenever
another fish, especially another male, enters the territory, it
is attacked. The attack, however, is rarely followed through
and it would be more accurate to say the intruder is threatened. The "threat stance" is very distinctive: the rays of the
dorsal fin bristle, the mouth opens as if to bite, the head
points down, and the body threshes a bit as if the stickleback were trying to dig into the sand. The ventral fins may
also flare. The intruder generally does not persist, but
moves off.

The male then proceeds to build his nest. Once it is
finished, the intensity and contrast of his courtship colors
become even more pronounced and he actively patrols his
territory as if on parade. Meantime the female, who takes
no part in building the nest, has taken on a brilliant silvery
hue and her abdomen swells with developing eggs. The females travel in groups, often passing near the males' territory. If a male happens to be ready to mate at that time, he
performs a sort of dance. It consists of a series of incomplete
loops in which he feigns a retreat from a female and then
suddenly veers back toward her, mouth open. Most of the
females seem frightened and move off, but those who have
reached maturity are not so timid; they turn toward the
male, head pointing up. At this point the male circles a
female, then swims toward the nest, followed by the
female. The male pokes at the entrance to the nest, as if
pointing it out to the female, who finally enters the nest,
her head protruding from one end and her tail from the
other. Then with his snout the male rubs her feverishly at
the base of the tail; after a moment, the female raises her
tail and deposits her eggs, then leaves the nest. The male in

turn enters the nest and fertilizes the eggs. He then chases the female out of his territory, repairs the nest, and takes care of the eggs, watching over them until they hatch (Tinbergen, 1951).

b. *The herring gull* (*Larus argentatus*). Herring gulls, which live in groups in fall and winter, become more or less solitary in the spring when they go to nest in the dunes. The flock then splits up into pairs, but there remains a group of unmated birds, and it is within the framework of this group that new pairs continue to be formed, the female taking the initiative in this case. She approaches the male, assuming a very distinctive posture: neck outstretched, feet flexed, she slowly circles the male of her choice. The male may then react in one of two ways: either he begins to attack other males or, after emitting a long cry of alarm, he goes off with the female. She then begins to beg him for food by shaking her head in a curious way, and the male regurgitates a little and gives it to her. Male and female stake out a territory together and both participate in building the nest. They copulate once or twice a day after a special ceremony: one of the birds shakes its head as if begging for food, then the male stretches out his neck, jumps in the air, and mounts the female.

Fights with other males may occur at any time, but while the nest is under construction the male is particularly intolerant of intruders. Actually, as with the stickleback, real battles are rare among herring gulls; it is more a matter of intimidation. At first the male confines himself to stretching out his neck, pointing his bill downward, and sometimes spreading his wings. He will then advance on the intruder in a very determined fashion, all muscles taut. If that is not enough, he goes up very close and pecks furiously at the ground, tearing up tufts of grass and flinging them about.

Most of the time the intruder will go away without re-
sisting.

    c. *The grayling butterfly* (*Eumenis semele*). These but-
terflies emerge in July and pass their time gathering nectar
in groups of 5 to 10 or even more. Soon the male will begin
to take up a position on the ground or on a branch and fly
up after any butterfly that passes by. If it happens to be a
female, she reacts to his pursuit by fluttering to the ground
and the male alights behind her. He then approaches and
confronts her. Unless she responds by beating her wings
(which means she is not ready to mate and that suffices to
discourage the male), he proceeds with the courtship
proper. He begins by rapidly beating his wings for a
moment and then, folding them back so as to display the
beautiful ocelli on the underface, he waves his antennae
and rhythmically opens and closes the front edges of his
wings for several seconds or a minute. He then spreads his
forewings wide, bows forward, and closes them over the female's antennae; this movement is very rapid, lasting
scarcely a second. Finally the male refolds his wings and
quickly passes behind the female to copulate.

2. Mechanisms of Social Interaction
before and during Copulation;
Synchronization of the Maturation Processes

    Clearly, males and females must be ready to mate at
the same time. The season, particularly the change in the
length of the days, gradually induces estrus through the in-
termediary of various hormonal circuits. But closer attune-
ment may occur as a result of the sexes coming together; for
example, the presence of a male, even caged, will activate
maturation of a female. In pigeons, the milky secretion of

the male's crop is not produced in isolation; a conspecific must be present, male or female, or even just the pigeon's own reflection in a mirror.

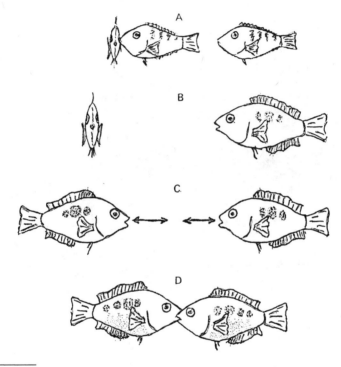

Figure 24. Sequence of social interaction between cichlids, *Etroplus maculatus*. One fish first approaches the other laterally (A and B), then they oscillate back and forth facing each other (C) and finally grasp each other by the mouth (D). (After Wymann and Ward, 1973.)

a. *"Persuasion" and "appeasement" rituals* (Tinbergen). On the one hand, males manifest very belligerent tendencies, while, on the other hand, females show a strong antipathy to being touched, not only by an experimenter but even by a congener. This aversion to contact exists in all species to a greater or lesser degree and is even evident in certain domestic animals. Taken together, these tendencies

Figure 25. Various forms of encounter and buccal contact between ground squirrels. a: the animal on the left takes the initiative in buccal contact. b: the animal on the left presses the cheek pouches of the other who has just taken some food that the first wants. c: buccal contact is refused. d: a subordinate animal, on the right, cautiously approaches a dominant one. (After Steiner, 1970.)

are the cause of much mutual suspicion which would hinder sexual approach were it not for one of the partners showing a little "diplomacy."

Although different coloration in the sexes usually serves to prevent the male from attacking the female, in some species the two sexes look just alike. In that case, when the male attacks, the female will avoid a "masculine response"; she assumes the typical female posture which often resembles an infantile attitude (such as the begging behavior that characterizes young gulls and reappears in the adult female). Prior appeasement is of the utmost importance in many species of spiders where the male is much smaller than the female and runs a great risk of being eaten. In some instances, he offers her a gift of prey; in others, keeping to the edge of the spider web, he gives it a series of shakes according to a special code, which provokes a kind of momentary catalepsy in the female. The male must then copulate and depart in haste under pain of being devoured.

"Appeasement" sometimes operates in very odd ways. In cockroaches, for example, the male raises his wings to reveal a gland located at their base; the female then approaches to lick the glandular secretion, whereupon the male seizes this opportunity to copulate. In certain crickets, the male stretches out his legs, which bristle with long spines; when the female approaches to break off the tips of these spines where an attractive secretion forms, the male profits from the occasion to achieve his ends.

b. *Symbolic appeasement.* Sexual approach involves some of the most bizarre rituals in the animal world. For example, when a male tern wants to mate, this sea bird begins by catching a fish which he takes to the female, who then holds it in her bill, without eating it, throughout copulation. In *Empis*, a minuscule predatory fly, the male offers the female a victim encased in a ball of silk, which the female proceeds to rotate between her legs during copulation without eating the prey. In other closely related species, the male simply gives the female an empty silken balloon! *Nuptial cannibalism* is a ritual practice among certain species of insects. In the praying mantis, no sooner does the male mount the female than she decapitates him. This by no means interrupts copulation. Quite the contrary, *consummation is possible only in the absence of the male's brain*, which seems to act as an inhibitor and must first be devoured by the female. There is a species of Hemiptera (a kind of bedbug) that copulates face to face, which is rather rare. During the copulation the female inserts her hornlike proboscis into the buccal cavity of the male and, piercing the wall of his esophagus, she secretes enzymes that completely dissolve all his internal organs. She then siphons off the liquefied matter and this terminates copulation which has been going on the whole time!

c. *Mutual orientation.* There are innumerable mechanisms that help the sexes locate each other. Song should undoubtedly be ranked at the top of the list as the most differentiated. It has been investigated in particularly great detail in grasshoppers, which are equipped with hearing organs that are highly developed but located in rather strange places to our way of thinking: on the sides of the thorax in Acrididae (short-horned grasshoppers) and on the anterior tibia in the Phasgonuridae (long-horned grasshoppers). Insect songs are often, but not always, the province of the male. Several varieties can be distinguished. One, which might be characterized as a "rivalry song," is used to respond to other males. The "love song," as it is called, has a strong attraction for females, who can readily differentiate the call of their own species; it helps them find their way to males they cannot see through the tall field grass. In an early experiment with crickets, Regen (1924) even succeeded in attracting a female to a loudspeaker that was transmitting the song of a male from another room. Regarding the love song of birds, a male that has not yet found a mate sings particularly intensely, but he ceases as soon as a female appears on the scene. Like grasshoppers, birds use a wide variety of songs and calls for warning, intimidation, etc.

Smell is the means of attraction most widely used by insects. In the large hawk moth (Sphingidae), the female attracts the male through an odorous glandular secretion. It is the gland alone, not the female as a whole, that interests the male; for instance, if the gland is removed and placed on a glass slide, he will try to copulate with it.

3. Sexual Behavior in Animals and Man[1]

While we have no intention of comparing animals and man in all areas, which would far exceed the scope of ethology, one must nevertheless admit that the area of sexual behavior is particularly appropriate for such a comparison. No one, in fact, has ever denied that in this area we come close to our animal roots.

a. *Variations in courtship. In man,* some males and a great many females still regard any form of sexual relations outside of coitus proper as perverted and distasteful, at least in civilized societies. These findings, which come from the Kinsey report, are inferred for a large part of American society and may hence be considered a result of the culture and, in women especially, of Puritanism. It is curious, however, that this extremely restrictive attitude toward preliminary activities is also found in primitive people such

[1] After Beach.

as the Kwoma, while others, such as the Trobriand, may devote hours to foreplay.

In mammals other than man, we find comparable variations. In certain cases preliminaries lasting several days are necessary before mating can take place. The proximity of a member of the opposite sex can, of course, hasten the development of receptivity, as in foxes for example. In chimpanzees, Yerkes and Elder (1936) observed that the presence of a male near a caged female is sufficient to induce the characteristic congestion of the genitals, sometimes within only a few hours. The male or the female or both may require a period of prior familiarization. Male porcupines will not mate in captivity unless they have been kept for several days in the immediate vicinity of the females' cage. Females, on the contrary, are immediately receptive to all males as soon as they come in heat.

They are also *individual differences*. Certain dogs pay prolonged court to a female, while others mate almost instantly. It is the same among anthropoids: a lengthy courtship with some and with others none.

b. *Seduction of the partner or precopulatory rituals. Sources of attraction*. In human societies, it is impossible to isolate any universal criteria for the sexual attractiveness of women. It has been observed that in a great many primitive societies an attractive woman must be plump; it is rare to find a predilection for slender women as in our society, a preference which is, moreover, recent (one only need think of Rubens). Often interest is focused on the head and face, especially the eyes. Among primitive peoples, little attention is paid to the hair, or to the nose or lips. In certain cultures the breasts should be firm and hemispherical; frequently, however, sometimes even in neighboring cultures, pendulous breasts are considered more esthetic. As for the physical attributes that women seek in men, they are as little known in any culture as they are in our own. In all societies, lack of bodily cleanliness is considered repulsive, as are signs of skin disease.

Also in all cultures, it is forbidden for a woman to display her vulva, even when both sexes go naked. There is a whole code of behavior that dictates how a woman should sit or lie in order to conceal her genital parts; showing them is regarded as a direct invitation to intercourse. Breasts and buttocks, however, which are often bare, provoke no advances on the part of men. On the other hand, in many societies the men go entirely naked and make no effort whatever to hide their genitalia.

A curious trait common to quite a few cultures is the esthetic value placed on large labia, which should be as fully developed as possible. Young girls sometimes manipulate the labia in order to elongate them. Among the Hottentots, this physical feature constitutes a racial characteristic (the Hottentot apron).

In animals, and especially in anthropoids, the erogenous zones are likely to interest the majority of males, an example being the enormously swollen genitals of the female chimpanzee in heat; these parts are handled and licked by the males. In chimpanzees in particular, some males are often much more selective than others. Yerkes and Elder (1936) observed instances when a male would refuse a receptive female and wait for another that was not in sight at the moment. The source of such differences in attraction is not known. One factor is certainly the intensity of heat. At the beginning of estrus, it seems that the female copulates mostly with lower-ranking males; upon reaching the maximum of receptivity, she mates with the leader; and toward the end of her period of sexual activity, she returns to the subordinate males.

Finally, some males seem to develop such marked physical preferences that they will attempt to change a female to suit their desires. Tinklepaugh (1932) reports the case of a pair of macaques who had been partners for several years but twice were temporarily separated. During separation the long facial hairs of the female grew out in a peculiar way. Shortly after they were reunited, the male proceeded to pull out and bite off all the long hairs that had grown on the female's cheeks, stepping back from time to time as if to judge the effect; the physical aspect of the female was completely changed.

Female apes and monkeys also show certain preferences: it seems that they single out the male with the longest penis. In addition, the most selective males in the choice of a partner are less attractive to females than those who are not so finicky. In chimpanzees, the male that is most dominating and roughest with females attracts them the least; in other species, however, it is exactly the opposite, for example, among macaques (Carpenter, 1942). There are similar indications of selectiveness in horses, cattle, and dogs.

Invitation to coitus. If without exception custom forbids women to expose the vulva, even among people who go naked, it is because in all cultures exhibition of this organ constitutes a direct sexual invitation. Among the Lesu, the Dahomeans, and the Kourtatchi, a woman who desires a man shows him her vulva with almost as much insistence as is found among simians. Perfumes and music are often utilized with very explicit erotic intentions. Very frequently both sexes perfume themselves, even more frequently than the women alone. This is to mask body odors and draw attention to personal cleanliness which is so important; in addition, certain fragrances are reputed to attract or excite men or women respectively. Serenading young girls with various musical instruments (some seem to be reserved for this purpose) is far from being a peculiarly Spanish or Italian custom. On the contrary, it is found among a great many people all over the world. In the western Carolines, the men sing before and after intercourse. Certain songs sung at midnight are said to draw the young girl toward the young man by magic.

Gifts often accompany sexual solicitation and are presented sometimes before but more often after coitus. This is true not only among primitive peoples but also in our modern societies, quite apart from any form of prostitution (gifts of flowers or candy, for example).

Last but not least to be sure, *speech* is used in sexual solicitation. In a very high proportion of societies, invitation to coitus is reserved to the man; it is forbidden or shameful for a woman to take this initiative. Among the Kwoma, however, it is the woman who must make the first advances; among the Trobriand, the Lesu, and the Kourtatchi, the initiative may originate with either sex and be considered normal.

Almost exact parallels can be drawn between what occurs in man and in animals. As for *smell*, myriad examples among mammals show its importance in triggering the sexual appetite of the male (for it is usually a matter of odors emitted by the female). There are no examples, however, of mammals other than man perfuming themselves with artificial fragrances, which animals generally seem to abhor.

As regards *song*, innumerable examples exist of its role in precoital ritual, though admittedly such examples are confined almost exclusively to birds and certain insects such as grasshoppers. Birds display a great variety of calls, some being invitations to the female, others threats aimed at possible rivals.

Nevertheless, we are all familiar with the piercing and cacaphonic calls of mammals when the males are in rut or the females in heat, as in deer, pigs, cats, etc. Many of these calls are heard only at mating time. A number of primates also produce a special sound by clicking the tongue; this signal is directed at the prospective partner and intended solely as a sexual invitation. Male baboons produce a peculiar clacking noise by pressing the tongue against the upper teeth and rapidly withdrawing it. Bingham (1928) produced this sound in the presence of male baboons and observed them go immediately into erection and seek out a female in order to copulate. During coitus, baboons emit a series of muffled grunts which seem to cause sexual arousal in other baboons within earshot.

Concerning gifts, many birds have a characteristic ritual which closely resembles a gift-giving ceremony: however, one cannot be sure that the phenomenon in question is identical to that observed in men in similar circumstances (see p. 101). Unfortunately we know of nothing analogous in simians, especially anthropoids. At most is seems that the male chimpanzee is particularly attentive to his chosen female and invites her to take the best morsels of food. The observations that have been made in this connection are quite old, however, and new research would seem to be called for.

As to *sexual initiative*, that is, which sex makes the first advances, it is commonly believed to be the male, but that is far from a general rule. In chimpanzees, when the female's genital region is congested, it is she who takes the initiative 85% of the time (Yerkes and Elder, 1936). But some females never take the initiative, even at the height of estrus. On the other hand, the male is sometimes slow to respond: the female will then repeatedly exhibit her genital organs and emit cries which end up sounding like shrieks of rage. This occurs most often when the male has copulated several times with a consequent sharp drop in excitability.

It would seem that in no species of mammal does the initiative fall solely to the male or female. By and large, it involves both sexes, and if one shows greater initiative, it is often the female.

The role of grooming in foreplay and sexual stimulation. In both men and animals grooming plays an important role in preparing for love-making. Among the Siriono, men as well as women spend hours combing their hair, delousing themselves, removing thorns and ticks from their skin, and rubbing each other with red dye from the roucou tree. Such behavior seems to be characteristic of primates, so much so that some primatologists tend to regard grooming in apes and monkeys as a secondary form of sexual activity, and, it must be added, it is indeed very often associated with mating. In macaques, delousing is observed to be particularly intense during intervals between repeated penetration prior to ejaculation. According to Yerkes (1948), the less pleasing a male chimpanzee is to a female, the more time he will spend trying to delouse her. In other animals besides anthropoids, the male and female sometimes pass long hours licking each other before mating. In any event, there seems little doubt that women *groom themselves* with a view to arousing their partners, in the immediate or more distant future. Other primates, however, primarily *groom each other*, not themselves. To find personal grooming associated with mating, one must turn to the Australian bowerbird; after building a courtship arbor, the male crushes blue berries and paints his breast with them. Perhaps the ritual dances of chimpanzees to which Yerkes alludes are related to such practices, for before the dance males or females may try to daub a little mud on their heads or wrap vines around their necks.

One form of sexual stimulation is essentially unknown in animals, *stimulation of the female breast*. There are also a few human societies where this type of stimulation is unknown, for example, among the Kwoma, the Siriono, and the Kwakiutl. In apes and monkeys, stimulation of the female's teats—for they do not have breasts as such—does not occur.

Kissing, with the mouth closed or open, is far from being universally practiced or even known in all human groups. It is regarded by many Asiatic peoples as repugnant (the Japanese for example) or ridiculous (the Thonga). It is unknown in Bali; there one finds facial friction, which led Europeans to believe that "rubbing noses" was a substitute for kissing.

In animals, gestures are observed that are similar to or even the same as kissing. Young chimpanzees often play by pressing their lips to those of a partner, then turn immediately to stimulating their sexual organs. A male may take the lips of a female in his own and suck vigorously on them during copulation. In howler monkeys, a female wanting to mate solicits a partner by sticking out her tongue; if a male is slow to respond, she may lick his face and hands. Lastly, male and female elephants put their trunks in their partners' mouths before coitus.

Manual or oral stimulation of genitalia. Although stimulation of the woman is practiced by almost all human groups, manual or oral stimulation of the male genitalia by the woman is decidedly less frequent. But there are cultures (e.g., the Trobriand, Tikopia, Wogeo) in which men are forbidden to touch their own genitals and the female organs as well; in such cases, it is the woman who takes the initiative in fondling and guides the penis to her vagina.

Anthropoids also practice such stimulation. It is very common among chimpanzees and with them, too, the male is the active partner, the female merely presenting him her genital region with insistence. Excitation of both partners then increases appreciably and copulation follows soon thereafter. When two monkeys have been living in the same cage for a long time, preliminary manipulation is essential for copulation, as if its purpose were to combat the effects of familiarity. It is rare for the female to manipulate or pull the male's penis to induce erection or attempt to insert it in her vagina, but it does occur sometimes. Fellatio is equally rare.

Painful stimulation. In Western cultures it is unusual, but not unknown, for painful stimulation to be associated with love-making, leaving aside, of course, various perversions of a sadistic nature. It is much more frequent, however, and even quasi-normal among certain peoples such as the Siriono and the Trobriand Islanders. Among the Siriono in particular, sexual foreplay consists of scratching and pinching each other, and even attacking the partner's genitals; in addition, male and female alike bite deeply into their partner's neck and chest during the sexual act. Among the Trobriand, it is the woman who is the more aggressive; she literally lacerates her partner's back with her fingernails.

This sort of painful sexual stimulation is not at all unusual in animals. Male shrews, bats, rabbits, and cats bite deeply into the female's neck during mating. In mink, marten, and sable, the bite may be so deep that the long canines meet, piercing right through the neck skin. The female thrashes about and tries to escape; this battle goes on for a long time and only when the female tires and struggles less does the male attempt to penetrate. Many males refuse to mate with a female that is too compliant; if nevertheless they do mount such a female, fertilization is rare. It seems that this violent battle may be essential for ovulation, for when a female mink fights with a male for a while, she may ovulate without copulating.

Similar phenomena occur in primates. When sexually aroused, male baboons will alternately chase, bite, and copulate with the female. Almost all macaques do the same, and it seems that certain females become receptive only after having been pursued and bitten by a male. A female in heat makes persistent advances to a male by displaying her hindquarters to him. She is often seriously bitten, but her only reaction is to keep after the male that has wounded her, with the result that after 9 or 10 days in heat, the female macaque is covered with scars and wounds inflicted by her partners. Among chimpanzees, however, copulation is accompanied by neither bites nor other injuries and during estrus the female even becomes dominant.

Finally, we should like to add an interesting, though hard to interpret, observation: societies in which the sex act is accompanied by painful practices are permissive vis-à-vis children and adolescents, from both the sexual and other standpoints. There are also permissive societies in which pain does not enter into precoital or coital relations, but they are rare.

c. *Time and place. The preferred place.* Little can be said about such preferences in Western culture, except that generally coitus is performed in seclusion, away from the presence of others. In primitive societies, too, coitus takes place away from other members of the group. This is particularly noticeable in the numerous societies where people live communally in a single, sometimes enormous, lodge. In that case, intercourse is performed outdoors in the bush. In other societies where each couple has a hut, preferences vary: the Hopi believe that to be legitimate coitus must take place in the house; conversely, the Gond and the Yurok believe intercourse must be performed outdoors or else one risks losing one's wealth (which is stored in the house).

This need for intimacy does not appear to be shared by the majority of animals; on the contrary, it seems that the presence of congeners heightens sexual excitation. Almost all animals, however, manifest very strong preferences for a specific location. In fact these preferences are so strong, particularly in birds, that they may result in *inhibiting mating entirely if the mating site is unavailable.* In sea birds, for example, mating and nesting take place only on certain rocks, always the same, where immense flocks of birds congregate during the mating season. Nesting sites are quickly occupied, with the result that latecomers wait on neighboring rocks only a few meters away until a site becomes free, *without mating* or nesting in the meantime. Among mammals such preferences are less marked, but it seems that in many cases male animals are attracted to a place where mating has successfully taken place. They become in a way conditioned to it. Females do not manifest such preferences.

The preferred time. In Western culture sexual relations usually take place at night, but this is doubtless attributable to the fact that the day is devoted to work. Still unexplained, however, is the preference of men for light during intercourse, while women prefer the dark. In the primitive world, a great many societies have sexual relations both in the daytime and at night, but at least as many only at night. Some, such as the Siriono, prefer the late afternoon, after the bath. Certain others (the Chenchu) choose daytime exclusively because they believe that a child conceived in darkness will be born blind.

In animals it all depends on their activity cycle. For example, all anthropoids sleep at night and are active only during the day; hence it follows that they mate in the daytime. The opposite is the case among nocturnal animals such as rats.

d. *Frequency of coitus.* In human societies nothing is more variable or subject to more taboos than sexual relations. Widows and widowers must often abstain from intercourse for a period ranging from several days to several months or even years, the more severe restrictions usually being imposed on widows. For men, sexual abstinence is also compulsory at certain times of great moment, such as before the hunt, during fish curing, before or after war, etc.

Apart from these periods, the frequency is highly variable, often once a day, sometimes (among the Keraki) only once a week. The Arunta say that there is nothing exceptional about having intercourse five times a night. According to the Kinsey report, the most usual average frequency in civilized societies is from two to three times a week, but with substantial individual variations. Frequency declines very sharply with age, but never falls to zero for all individuals, even at a very advanced age.

In animals such as anthropoids, copulation is limited to the period of rut, man being the only primate in which sexual activity is not seasonal. At the beginning of the mating period in macaques, males may experience up to three ejaculations a day, but their sexual activity declines very rapidly. After only a few days, the female makes incessant advances, shrieking and exhibiting her congested genitals, but the male may be almost totally unresponsive. There are no examples of continual sexual activity as in man. In this connection it has been said, not without humor, that if man is the animal with the biggest brain, he is also the one with the biggest penis in proportion to his size. In animals other than man, it seems that the female's receptivity and sexual desire *when in heat* far exceed the capabilities of a single male.

e. *Types of sexual relationships.* In man, societies in which monogamy is considered the only lawful union are exceptional among primitive peoples. A whole range of sexual associations, including polygamy, is allowed, but nevertheless monogamy is the lot of most individuals for two reasons:

1. There are very few men among primitive peoples who can manage to feed several women; only the chiefs and wealthiest men can afford to do so.

2. Restrictions in another connection are extremely rigorous, namely, exogamic prohibitions, which forbid a man to marry women from a certain clan or which extend the concept of incest to distant relatives, with the result that in some instances more than half the available women are strictly taboo.

The fact of the matter is that polygamy is very often accepted in theory, but only rarely is it possible in practice. It should be added that polygamy is more widespread than polyandry, which is found in only a very few societies, one example being the Toda, where a woman marries her husband's younger brothers.

In animals, it is rather difficult to determine the types of sexual associations that prevail among anthropoids. A key factor would seem to be dominance, which may be very rigid (macaques, baboons), rather vague (chimpanzees), or almost nonexistent, at least from the sexual standpoint (gorillas). Gibbons provide one example, probably rare, of the monogamous family. Their small groups include only one male and one female adult, apparently united for life or at least for a very long time; the young seem to be chased out of the family group as soon as they reach sexual maturity. This example of monogamy is particularly noteworthy because it is so unusual. Among other mammals, almost the only example that can be cited is wolves, whose social organization on the whole resembles that of gibbons. In wolves, however, young adults remain in the group headed by the monogamous couple; they help with the training of the cubs and never or almost never copulate. These findings, which seem quite reliable, nevertheless call for further verification; for, if they are confirmed, in wolves we have the only instance in which mammalian social organization approaches the social organization of invertebrates, i.e., that of bees or ants, with a very small number of reproductive individuals while the others, far more numerous, remain sterile.

It is possible that permanent monogamous unions similar to that of wolves may exist among foxes. This hypothesis is based on the experience of breeders. Their concern is to mate the females with the males having the most beautiful fur, which means that a given male must fecundate several vixens during the brief rutting period. The usual behavior of foxes runs counter to this procedure, for once a male has been able to form a sexual bond with a female, he is unresponsive to all other females. Therefore, the males must be "deconditioned," so to speak, by placing them as young adults with receptive females for only a few hours at a time. In that case, the males become trained to fecundate a new female every day. If, however, a young male is left with the same female overnight, it will be difficult or impossible to get him to respond to another female for the rest of the season. A vixen in heat, however, will accept any male whatever.

Zuckerman (1932) studied a colony of captive baboons (a method of studying animals that has been called into question by modern ethology, we might add). Out of 25 males and as many females, 5 of the males mated with all of the females while the 20 other males had no partners at all. It is social dominance and physical prowess that determine the formation of polygamous unions among baboons, no different from what is found in primitive or even civilized human societies.

It should be pointed out here that *examples of monogamy are frequent in birds*, where the partners may remain the same for several years and probably for life. There are, however, also cases of monogamy limited to a single season and of polygamy as well. It depends on the species.

f. *Sexual taboos among animals and the question of incest.* Prohibition against incest is very general as regards parents being forbidden to form unions with their own descendants. This rule applies to everyone except sometimes chiefs as among the Azande. The same holds true for the prohibition of sexual relations between brothers and sisters. It is also a general rule except that such relations may be obligatory in very rare cases and for rulers only, for example Azande chiefs, Inca rulers, or the pharaohs of ancient Egypt.

Japanese research on macaques clearly shows that a family relationship exists among monkeys which lasts at least for several years. It is particularly evident in the monkeys' grooming behavior (they prefer to be deloused by their brothers or sisters) and in fights (a mother may defend her son for several years after he has left her). It also seems that *monkeys do not copulate with their mothers*. It should be added that according to Gabler (see Lorenz, 1952) this "incest taboo" also exists among wild geese. Sokolovsky's (1908) early observations of a group of chimpanzees showed that the dominant adult male availed himself of all the females but was vigilant in cutting short any attempts of the younger males to copulate. For their part, the females seemed quite receptive to any male at all whenever the despot relaxed his guard. Carpenter (1940) believes it is similar among gibbons: the dominant male's jealousy chases all the mature young males out of the group, his partner's jealousy working in the same way with respect to the young females. It would be most interesting to verify whether this is really how the rejection of young adults occurs.

Among baboons, a male possessing a harem keeps his females under constant surveillance to prevent them from having relations with another male. Moreover, celibate baboons rarely try their luck; if they do, it is a sign that the tyrant's prestige is on the decline. Females, on the contrary, take advantage of the slightest distraction of their master in order to copulate. When a female is surprised by the leader during copulation, she can avoid punishment by a wily little game; she immediately interrupts the sex act and runs to her master, offering herself eagerly to him and at the same time grimacing at her seducer. It is then the intruder who is sharply corrected, the female escaping punishment. Among macaques, even when the male is sated, which happens very quickly, he will still try to prevent his mate from coming in contact with other males, but to no avail (Carpenter).

g. *Homosexuality*. In modern societies homosexuality is not rare. The Kinsey report maintains that out of a sample of 5,000 Americans 37% had had homosexual relations with orgasm after puberty; however, we have seen that there are grounds for reservations regarding Kinsey's sampling. Among his subjects only a small number went so far as anal or interfemural copulation, the greatest number stopping at mutual manual masturbation. It seems that 26% of the women studied by Hamilton (see Beach, 1965) had had homosexual contacts; for the most part they practiced cunnilingus and, to a lesser extent, manual masturbation of the clitoris.

Among primitive peoples, one finds a whole range of attitudes, from very strict prohibitions against homosexuality, even under pain of death (among the ancient Hebrews, and in our time among the Rwala Bedouins), to more or less broad tolerance. Societies that are "tolerant" on this point are the only ones where it is possible to obtain information in any detail. In certain cases homosexuality is ritualized in a way, as with the shamans of Siberia, for example. The shaman of the Chukchee is customarily homosexual and enters another man's household as his "spouse," though the man already has a wife and children. Often the shaman himself is also a husband and father. He is believed to have been transformed by a supernatural power and is highly respected. Among the Crow Indians some men dress like women and live alone; other men and adolescents frequent them and they practice fellatio on their visitors. In addition, sodomy is very frequently an initiation custom, being one of the trials the young men must undergo and in turn inflict on others after being initiated.

Homosexuality must also exist among women in primitive societies, but it is much rarer than it is among men or, more probably, it attracts little attention. A special instrument is quite often used by women (Nama, Azande).

In animals, homosexuality is very frequently found in primates. In a case reported by Hamilton, an adult macaque formed such an attachment to a young male, a relationship marked by frequent acts of sodomy and social protection which caused the young animal to rise in the hierarchy. When the young male was absent but females were present, the adult male mated readily with them and behaved heterosexually in every way. But as soon as he was reunited with the young male, his homosexual practices resumed, without, however, putting an end to his heterosexual activity. Still, as the Japanese have noted, it should not be thought that coitus between macaques, whether of the same or the opposite sex, is always genuine; very often it is only a matter of one of the partners assuming an attitude of submission while the other merely simulates copulation.

Homosexual behavior has also been observed in male chimpanzees and baboons. It is found among celibate baboons when the dominant males have taken all the available females. They indulge in all sorts of practices ranging from anal coitus to mutual masturbation, either partner being capable of adopting the male or female role in rapid alternation. In fact, homosexual coitus is so frequent among young primates that some authors postulate that at first there is no sexual discrimination and it develops only gradually. Sometimes it remains incomplete throughout life, for many monkeys will copulate with a female and then within a few minutes perform anal coitus with a young male.

On the other hand, one must not overlook the important role of sexual posture (vis-à-vis animals of the same or opposite sex) in the social hierarchy of apes and monkeys. In fact, there is a tendency for female behavior to develop in socially inferior males. Maslow (1940) maintains that the behavior of a dominant animal is always masculine in nature, even if the dominant one happens to be a female. Furthermore, a subordinate's gesture of sexual presentation is not necessarily followed by real or simulated coitus with the dominant animal; it is frequently followed by permission to take some food or indulge in some activity which the dominant animal would not otherwise tolerate.

Homosexuality among female primates is much rarer; it is doubtful, in fact, whether it even exists. In other mammals, however, homosexuality among females is more or less normal. Heifers in heat readily mount other heifers; this *male* gesture is a sure sign of *female* receptivity on the part of the heifer in question. This phenomenon illustrates the ambivalence of the sex hormones, which in reality act to raise general excitation to a certain pitch. The male, on the other hand, has female components in his panoply of sexual behavior, just as the female has male components. It is a question of balance, with one set of components usually outweighing the other. At the beginning of the sexual cycle, one component or the other may, under the influence of intense hormonal discharges, come to the fore and trigger (in a way that is reversible) one type of behavior or the other.

Among males of species other than primates nothing is more common than homosexuality. Or then again, one of the sexes may temporarily assume the role of the other, as in the case of heifers. For example, when female rats are confronted with a lazy male who is unresponsive, the females may "mount" him like males, while the sluggish male displays the characteristic lordosis of a female. This treatment excites him, it should be added, since very soon thereafter he can perform as a male.

h. *Relations with other species*. In our civilized societies sexual relations between man and animal are not unknown during adolescence. The animal involved is generally
a male or female dog. Some dogs can be trained to copulate
with a human being. What is most singular is the affective
bond that then develops *on both sides*, man for animal and
animal for man; in a male animal it is accompanied by
great jealousy vis-à-vis men.

Among primitive peoples, copulation with animals,
called "bestiality," is normally absolutely forbidden,
usually under pain of death for both the man and the parti-
cipating animal. There are, however, some groups, such as
the Hopi and Masai, where the practice is common, and in
certain ancient religions (Phoenician) bestiality constituted
a ritual, which was the object of angry maledictions in the
book of Leviticus.

In animals, seeking sexual contact with any animal
whatever, even one of a quite unrelated species, is not unusual under the pressure of acute, unsatisfied sexual need.
Everyone has witnessed the attempts of young dogs to
copulate with the leg of their master. In this realm, too, are
the African stories, all highly suspect, of chimpanzees or
gorillas carrying off women. These tales are patently quite
unlikely (though not absolutely impossible) because they
imply an abnormally intense state of sexual need in animals
that always live in groups and consequently should easily be
able to find satisfaction. As for apes and monkeys *in
captivity*, there is no lack of examples of them trying to
copulate with man. Yerkes relates how he was the object of
the eager advances of a young female gorilla of formidable
strength, but he fails to tell how he avoided succumbing to
her charms. When different primates are kept in the same
cage, interspecific copulation occurs very readily, even

between distant species, for example between a male macaque and a female baboon.

i. *Masturbation.* Among young boys and adolescents masturbation is extremely widespread if not universal. It is quite probable that in the great majority of cases that is how the first orgasm is obtained. Later on masturbation becomes much less frequent and in many, but not all, adults it disappears completely (findings of the Kinsey report). Female masturbation seems to be found in far fewer subjects, a little more than 50%. Many women do not go as far as orgasm; on the other hand, some can reach orgasm only in this way and not through heterosexual relations.

Among primitive peoples, masturbation in men is generally forbidden or regarded as a ridiculous or disgusting act. In young boys, however, it is ordinarily tolerated or considered unimportant. One also finds a few reported cases of female masturbation, which is usually severely punished by the men if they find out. The Lesu are the only group in which there is apparently no sanction against female masturbation.

In animals masturbation is universally widespread, particularly in male primates who assume all sorts of positions and employ all sorts of objects for masturbation. What is interesting is that in animals, too, masturbation is much more frequent in males than in females; in fact, we know of very few well-documented cases of masturbation in females, whereas examples in males are innumerable and unequivocal. Apes and monkeys generally use their hands to stimulate their genital organs and certain primitive monkeys use their prehensile tails. Elephants employ their trunks.

CHAPTER 6

The Development of Affectivity

The term *affectivity* might suggest anthropomorphism were we not to define it objectively. According to Harlow (1958), affectivity is essentially synonymous with *attachment* and can be measured by time spent in bodily contact with a congener. It is scarcely necessary to point out possible parallels in man. Actually, we are dealing with a rather low level of behavior here, one where the higher cerebral mechanisms hardly come into play, except, that is, when man indulges in romanticizing these primitive tendencies and embroidering them with flowery language!

Affective tendencies may begin to develop very early, which has caused a number of investigators to try to find out whether *the organism in utero might not even then be influenced by various stresses that affect the mother.*

1. PRENATAL STRESS[1]

When a gestating mother has been subjected to various stresses and one wants to measure the effect on the offspring, one must, of course, eliminate the effect the disturbed mother might have on her young immediately after

[1] After Archer and Blackmann (1971).

125

birth. The type of care she gives her progeny might well reflect the emotional shocks she has undergone. This is why cross-fostering is essential: the young produced by an emotionally strained mother are transferred to a mother not under stress so that the care given them will be completely normal.

Almost all researchers agree *on the nature of the response to prenatal stress:* it results in decreased locomotion in new situations and retarded appearance of a series of reactions in various types of behavior tests. It is necessary, however, to take into consideration what type of *stock* the animals in question come from, whether it is active or inactive. De Fries and Weir (1964) found that prenatal stress decreases activity in the progeny of active strains and increases it in inactive strains.

a. *Different types of stress.* Surprising as it may seem, different types of stressful situations seen to act in the same way and produce highly analogous final results, irrespective of the nature of the situation—handling, anxiety-provoking avoidance behavior, electric shock, etc. This derives from the fact that all types of stress produce essentially the same changes in the hormonal system, and hormones constitute the only link between the mother and the unborn young. Some writers have even suggested, though with some exaggeration, that the major factor in prenatal stress is simply the removal of the mother from her cage in order to subject her to the experimental agent.

The time when stress is applied is evidently important; depending on the stage of gestation, stress may ultimately yield quite different results, but this aspect of prenatal stress has not been studied systematically.

b. *Effects on learning.* The progeny of mothers subjected to stress generally show an increase in the time necessary for learning and tend to make more errors. This is perhaps simply due to differences in emotivity which are obvious in other connections. Such is the case in avoidance behavior, where stressed subjects show more avoidance responses and shorter reaction times, which of course facilitates acquisition of avoidance behavior.

We could therefore conclude that prenatal stress increases the emotivity or "timidity" of the progeny, who are, it should be added, significantly more susceptible to gastric ulcers (so-called tension ulcers, occurring in animals that are briefly immobilized by being strapped on a table, even though they are subjected to no other treatment).

c. *The influence of postnatal life on the manifestation of reactions to prenatal stress.* It would indeed seem that *postnatal* handling reduces the emotivity caused by *prenatal* stress. But a factor that is often overlooked by writers and yet is very important happens to be the *rearing conditions*, whether the animal is raised in isolation or in a group. When handled, rats from groups are less emotive (less vocalization, for example) than rats raised in isolation. Isolation, then, increases the effects of prenatal stress.

Finally, of course, reactions resulting from prenatal stress diminish with age. Measurements should be made very early; rearing in isolation and postnatal manipulation both prolong the effect of prenatal stress considerably. Its most striking characteristic, though, as opposed to postnatal stress, is that *it is not lasting*.

2. Postnatal Disturbances

One of the greatest discoveries of behavioral science in recent years, one that had its beginning in the work of Denenberg (1962; Denenberg and Bell, 1960), is the ease with which the equilibrium of the newborn can be upset by stress from seemingly insignificant life experiences (stress upon the gestating mother was not studied until later). Handling consists of "gentling" or fondling newborn animals for a few minutes each day and then returning them to the litter. Such handling is enough, however, to induce several months later (which would correspond to several years in man), long after the handling has ceased, very marked behavioral differences as compared to control animals not so handled.

a. *Experiments before weaning.* The effect of early handling first becomes evident in the *learning faculties* of the adult. Levine et al. (1958) found that rats handled during the first three weeks of life are later superior to control animals that have not been handled. On the other hand, if the handling takes place from the eighth through the tenth weeks of life, one finds no effect later on. A weak electric shock can replace the handling (in certain instances, it even accelerates growth!). As for *emotivity* (measured by the frequency of defecation of subjects placed on an illuminated platform), rats and mice that have been handled seem less emotional; this is perhaps why they perform better in learning tests.

Neuroses may be induced experimentally in lambs and kids subjected to very early conditioning (electric shock following a signal). The animal will thrash about, pant, bleat, and defecate as soon as it hears the signal. Later, a genuine neurosis will appear at the age of two years, but, interestingly enough, *it will not develop if the animal was conditioned in the presence of its mother* (Liddell, 1954).

Cross-species adoption. This is a very interesting technique that has been too seldom used. To take an example, lambs raised by goats learn to fight by rearing up on their hind legs like kids and later they will display great fear when placed with animals of their own species. A curious finding of Blauwelt and Richmond (1960) is that kids or lambs raised by other species grow *faster* and are livelier than when raised by a mother of their own species!

Bottle feeding. Prolonged isolation and bottle feeding of young animals produces alterations in growth potential and emotivity. Most writers tend to agree that young bottle-fed lambs and kids are more active, become fattened earlier, and respond more vigorously; they are not susceptible to "trances," that is, to prolonged reflexive immobilization, as are normal lambs and kids. A bottle-fed female that was studied by Scott still showed behavioral anomalies after having been reunited with the flock for a year. She would not permit her lambs to suckle and paid little attention to them, scarcely reacting when she was separated from them. Bottle-fed males adjust with difficulty to the flock and are less aggressive when in rut.

Number of young in the litter. According to Seitz (1954), female rats give more maternal attention to offspring in small litters. Young from litters of 12 later manifest a much stronger tendency to hoard food than do young from litters of 6; they copulate more and weigh less.

Physiological consequences of early handling. The consequences of early handling are so great that *even the biochemistry of the brain and adrenal glands is modified.* Levine and Alpert (1959) have shown that the brains of animals that have been handled contain more cholesterol and their adrenal glands less ascorbic acid than normal. Mortality is also affected: it is lower in rats and higher in mice that have been handled. In addition, blood cholesterol is higher in fondled rats.

A puppy that is handled will show changes in heartbeat during the first two weeks. The increased rate that one finds would seem to be due either to diminished influence of the vagus nerve or to heightened sympathetic tonicity. The adrenal glands are also highly developed and there are indications of changes in lipid metabolism and in the brain as well. The encephalogram of a young puppy subjected to handling resembles more closely that of an adult animal.

b. *Experiments after weaning.* Most research after weaning involves modifications of the environment in which the animal is raised. The three main types of environment are normal, deprived, and enriched, but the meaning of these terms varies considerably depending on the writer. A normal environment is that which prevails under ordinary laboratory conditions, which are not very well defined. In a deprived environment, instead of being kept in wire cages, the animals are often kept in compartments with walls of frosted glass which hide the surroundings from view. Generally writers are in most accord about what constitutes an enriched environment: it is a rather large cage containing various objects, playthings, tunnels, inclined planes, etc.

Rats placed in an enriched environment after weaning are more successful in various types of learning. In this connection a comparison was made, however, between rats from a so-called "clever" strain and rats from a "dull" strain. When the dull strain was placed in an enriched environment, its subsequent results in maze tests were as good as those of the reportedly clever strain, and when the latter was placed in a deprived environment, it later proved to be no more successful in maze tests than the dull strain— which leads us to conclude that such *categorization of rats is valid only if they are raised in a specific environment.*

Sensory stimulation, especially if it is rich and varied,

also modifies the perceptual capacity of subject animals. If panels displaying geometric figures are hung in their cages, young rats raised under those conditions will later be more adept at differentiating various figures in discrimination experiments. Wolstenholme conducted an experiment in which kittens were raised in vertically or horizontally striped surroundings. Later, not only were the kittens exposed to horizontal stripes unsuccessful in learning vertical stimuli and vice versa, but in addition their retinal neurons showed marked atrophy in terms of the vertical or horizontal.

Sensory stimulation also affects *exploratory behavior.* In animals raised in a restricted environment, exploratory behavior increases as soon as they are placed in a free environment (Woods et al., 1960).

Socialization disorders in dogs. The relationship between dog and man does not develop as automatically as one might believe. It is in the first 13 weeks that this relationship is most easily established; after 14 weeks it becomes more difficult and young dogs raised with no human contact until that age are wild and unapproachable. On the other hand, a puppy taken at the age of 4 weeks to be raised with humans will later be unable to readjust to the kennel and will be impotent. In another connection, remembrance of a painful stimulus is achieved at 8 months and lasts a very long time. That is why great care must be taken when young dogs are vaccinated that they do not come to associate the pain of the vaccination with some commonly encountered situation; the great fear that some dogs have of entering a car derives from such an association.

"Hysterical" behavior is observed in dogs of small stature that are frequently petted and picked up by their masters (for being picked up is a sign of dominance among

dogs). As the owner replaces the dog's mother, any disturbance will cause aggressive behavior toward strangers, depression, anorexia, etc. When a fracture has been set, the attentions lavished on the patient may later cause the dog to feign a limp or refuse to use its leg. The simulated lameness would seem to be solely for the purpose of attracting the master's attention (compare the wiliness of the mother partridge that pretends to have a broken wing and thereby attracts enemies away from her nest).

In conclusion we should like to add that puppies that receive too much attention become entirely dependent upon man and show asocial behavior vis-à-vis other dogs.

3. DEPRIVATION EXPERIMENTS

Being deprived of the mother or companions at an early age and for prolonged periods causes very serious and sometimes irreversible consequences in most higher animals. This is also true of other forms of deprivation.

a. *Food deprivation.* Being deprived of food has far-reaching effects on behavior, even long after the subjects return to a normal diet. Rats, for example, generally store away food pellets once they are satiated, but investigators wondered what causes the considerable individual variations in this hoarding behavior. They discovered that such differences are directly related to *deprivation at an early age.* Food deprivation also affects social behavior; after a period of inanition, deprived subjects will fight much harder than control animals to carry off a food pellet. There would seem to be a critical period, however, and, according to Hunt (1941) at least, deprivation of food after 32 or even 24 weeks will have no effect on later behavior.

b. *Consequences of isolation.* An environment that is too monotonous and too poor in stimuli produces autism in isolated monkeys, a syndrome characterized by excessive timidity, stereotyped activity, and hyperactivity. In children (where the etiology is more obscure), the disorder is characterized by a more or less complete absence of verbal communication and an apparent lack of reaction to painful stimuli. A puppy isolated in a very small cage also shows this lack of sensitivity; for example, it will repeatedly approach a candle flame though it burns its nose.

Some species of animal are more sensitive than others to isolation, especially at certain critical periods. The electroencephalogram of an isolated animal shows constant arousal with low-amplitude wave activity, which closely resembles the encephalogram of an autistic child. Isolated animals are very nervous and sometimes merely opening the cage is enough to trigger convulsions. The stress of isolation can be mitigated by handling the animals for 5 or 10 seconds every day.

c. *Deprivation experiments with monkeys.* In a series of celebrated experiments on deprivation, Harlow (1958) isolated young rhesus macaques from birth, raising them on the bottle. Under these conditions they rapidly developed certain major and almost irreversible disorders. It was found that introduction of a "substitute mother" (simply a wire frame wrapped with a furlike cloth) gives a young monkey a sense of security and prevents at least many of the disorders due to deprivation of the mother.

Among the most serious anomalies that show up in isolated macaques, we should like to note those concerning sexual behavior which were observed by Harlow in 1962 and have also been found in other monkeys raised in complete isolation. Females will later manifest a strong aversion to submitting to the advances of a male. With males it is still more serious: they have frequent erections and even masturbate, but are incapable of mating. Females that eventually bear young become, in Harlow's words, "motherless mothers," whose behavior toward their offspring is very abnormal. Such a mother will push her baby away when it tries to suckle and even try to kill it by smashing its head against the floor of the cage.

d. *Behavioral disturbances and maternal care.* Although the behavior of motherless mothers toward their young is highly abnormal, it is curious to note that the behavior of their offspring is not abnormal to such a degree. Play seems to develop more slowly in them than in normal animals. Similarly, in young raised by an artificial mother, one finds a lack of aggressiveness, difficulty in joining in with a group, and delayed sexual development. If, however, motherless animals are afforded the company of other young congeners from birth, their development is perfectly normal in all respects. *The presence of congeners seems to replace completely the normal mother.* Repeated change of mother or repeated separation from the mother seems to cause no disturbances that can be detected at the age of one year.

But what happens if observations are carried out much later, at around 30 months? At that time subtle differences appear in deprived animals. Offspring of multiparous mothers are relatively more playful, more aggressive, and more agile. Young raised by motherless mothers are also more aggressive, but they enter into a group less readily. Their aggressiveness perhaps derives from the fact that their mothers are very rough with them. Instead of using vocal threats or menacing facial expressions, such mothers are quick to strike their offspring, who will in turn behave in the same way later on. Young that have been repeatedly separated from their mother are more fearful and less aggressive, while those who have had a frequent change of mothers are fearful and at the same time aggressive.

It is striking to note that all these disturbances do not show up until the onset of puberty. Sackett (1970) has undertaken a very detailed investigation of short- and long-term disturbances caused by isolation.

4. Innate Preferences in Monkeys

a. When a rhesus monkey enclosed in a compartment in the center of a hexagonal cage has a choice of any of the peripheral compartments where "stimulus monkeys" of the same or another species are chained, one finds that a young rhesus monkey that has been isolated from birth and never encountered monkeys save very young ones of its own age, which are quite different from adults, will tend to choose the company of a female rhesus and ignore animals of other species. This would indicate some sort of *innate disposition*, especially since young animals prefer a female to a male though they have never seen either one. On the other hand, adults who have undergone partial isolation (that is, who have been able to see but not touch other monkeys through the bars of their cages) show a preference for animals of their own sex. Rhesus monkeys that have been raised completely by humans prefer them to other monkeys; but those that have lived in a group, even though they may have been cared for by humans, prefer monkeys.

Therefore experience or lack of experience with a specific social stimulus within the first 30 days of life has a definite effect on preferences three or four years later. A total absence of social stimuli during the first six months permanently lowers the interest taken in social stimuli later on. A new social contact, however, immediately after the first month of life will influence the subsequent choice of a preferred stimulus. In short, for the rhesus monkey the experiences of the first six months of life are crucial for the formation of new attachments.

Preferences, though, may hinge on more subtle elements. For example, rhesus monkeys can be given a choice between monkeys raised like themselves or in a different way. In one experiment, the subject is allowed to play with each of the three monkeys from which it will later have to choose. In a follow-up experiment, however, the three monkeys are strangers to the subject. It happens that in either case monkeys will show a preference for congeners that have been raised like themselves (total isolation, partial isolation, or in a group). This also holds true in a community cage; when a new monkey is introduced, it will tend to draw close to those that were raised in a similar way.

b. *The effect of pictures.* Some groups of monkeys have been raised in total isolation except for being fed by human caretakers for the first two months. After that they no longer see anyone, either man or monkey. Instead they are exposed to slides or films showing monkeys occupied in all sorts of activities and also pictures of men, houses, landscapes, or geometric patterns. On certain days the experimenter, or the monkey himself, can change the picture by pressing a button. It has been found that such monkeys are more interested in the pictures showing monkeys in action than those showing something else. No picture elicited an aversive response before 80 days, but such reactions appeared soon thereafter, particularly with respect to pictures showing threatening behavior. Thus there exist *innate mechanisms that enable a monkey to recognize a form of behavior that it has never seen in a live animal.*

If, on the other hand, monkeys are exposed to geometric figures, a curious sex-linked difference shows up: females raised in isolation explore these figures much more than isolated males. In populations having lived in groups, quite the contrary, the males explore such figures much more than the females. A further distinction must be made between simple and complex figures; normal animals explore complex stimuli at greater length, whereas isolated ones prefer more simple stimuli.

5. ISOLATION AND PAIN

When one electrifies a tube which must be touched to obtain water, one finds that isolated subjects tolerate much stronger current than normal animals; the latter will cease seeking water much more quickly than the former. It seems that isolated animals learn avoidance behavior in connection with a painful stimulus much more slowly than normal animals. This recalls the lack of sensitivity in isolated puppies, mentioned earlier.

This relative insensitivity to pain explains the abnormal behavior of isolated monkeys when they are introduced into a normal group. They have no hesitation about attacking dominant animals who react by meting out sharp punishments, including deep bites; nevertheless, the isolated monkeys do not shrink from repeating the offense. Similarly, isolated animals frequently bite themselves very deeply, and will do so repeatedly, as if they forget that biting is associated with pain.

6. Are Isolated Monkeys Hypo- or Hyperemotional?

a. Several writers have maintained that animals raised in isolation are hyperemotional, basing their opinion on the fact when such animals are put in a new situation, especially when placed among congeners, they display fear, immobility, and general hypoactivity. It is postulated that this is due to the sudden contrast between the paucity of stimuli they were exposed to under their rearing conditions and the abundance of stimuli in the new situation.

Sackett and his colleagues, however, measured the amount of cortisol (the "stress hormone") in the blood of monkeys before and after putting them in a play cage where they met new comrades. Surprisingly, they found that *before* being put in the cage, the adrenal glands of isolated animals produce more cortisol than those of group-raised animals; *after* the introduction, however, they produce not more, but somewhat less. Isolated monkeys would then be hypo- rather than hyperemotional. It seems that in some way they inhibit the reception of new stimuli. Nevertheless their adrenal glands are just as capable of secreting high levels of cortisol as those of group animals. This can be demonstrated by injections of ACTH: the titration results are comparable for isolated and group animals.

b. *Isolation and learning.* In view of these disturbances, it is not surprising that isolated animals also show deficiencies when it comes to operant learning. Learning is much slower, particularly with regard to obtaining the first reaction. Afterwards, the cadence of learning is no different from that of normal animals. Extinction, too, is slower in isolated animals, which would seem to be connected with some difficulty in inhibiting responses that are no longer appropriate to the situation.

c. *Counteracting the effects of isolation*. In a series of experiments, Sackett (1970) compared the results obtained with different subjects: strictly isolated monkeys, isolated animals that had been shown pictures of monkeys, partially isolated animals that could see their fellows through the bars of their cages, and, lastly, monkeys that had been allowed to play with companions. It was found that visual stimulation by means of pictures is not successful in preventing isolation-linked deficiencies in *social behavior* (withdrawal, fearfulness, abnormal attacks, etc.). On the other hand, such stimulation can overcome deficiencies in *exploratory behavior* vis-à-vis new objects, deficiencies which are so pronounced in totally isolated animals. Furthermore, *females are much less susceptible than males* to the effects of isolation in the realm of social behavior.

7. Theories on Isolation

Several theories have been advanced to explain the catastrophic effects of isolation. One hypothesis is that of *structural atrophy*: being deprived of sensory excitation causes serious and permanent alterations in the corresponding sense organs; these changes are at the root of later behavioral deficiencies. Another somewhat similar hypothesis is that of *arrested development*: a number of physiological mechanisms are under- or undeveloped at the time of birth; to complete their development there must be a series of sensory stimulations at a certain critical period after birth. A third hypothesis is that of *learning deficiency*: organization and integration of excitation deriving from the environment must begin at a very early age if those processes are to function normally later on; if there is nothing to learn owing to a lack of excitation, the integration mechanism is not exercised and becomes permanently "rusty." For example, an animal's lack of social experience with a mother or congeners of the same age at a certain critical period can permanently lower its receptivity to excitation from members of the social community. Finally, a last hypothesis is that of *traumatic change*: traumatic effects are produced by too great a contrast between the stimuli to which the organism is accustomed in a restricted environment and the stimuli of a normal environment.

Sackett has also attempted to simulate the effects of social contact through cerebral stimulation (by means of electrodes implanted in the reticular substance, colliculi, hypothalamus, and the thalamic regions). In an animal totally deprived of sensory stimulation, electrical stimulation succeeds in partially preventing development of the isolation syndrome. This would tend to disprove the theory that the proper development of social functions requires *specific* stimulation: on the contrary, it would now seem that nonspecific stimulation is sufficient, though Sackett agrees that that still remains to be fully confirmed. It is also true that the presence of only one other monkey in the cage of an isolated subject is by and large sufficient to prevent the appearance of the syndrome of social deprivation.

Being deprived of contact with companions, as opposed to the mother, produces a whole series of anomalies, the most striking of which is self-aggression when exposed to a frightening object. Normal animals tend to direct their attacks against the object, but a deprived animal may bite itself so hard that it breaks a bone. In such cases even the projection of films showing other monkeys does not attenuate the effects of isolation.

8. The Theory of Affectional Systems of Harlow and Harlow

Harlow and Harlow (1965) conceive of affectional systems as a group of behavior patterns that establish and maintain social bonds in a group of animals of the same species. They distinguish:

1. The maternal affection of mother for child.

2. The affection of child for mother, which they also call the infantile affectional system.

3. The affection between playmates or companions of the same age class, which the Harlows think is probably the most important.

4. The heterosexual affectional system, which comprehends much more than periodic coitus. As Washburn and DeVore (1961) have shown, *the fact of sex is much more important than the act of sex* among primates. The fact of being of one sex or the other is of the greatest importance in social roles involving dominance, aggressiveness, play, care of children, rank, etc., *whether coitus occurs or not.*

5. There is also a *paternal affectional system.* The work of Suomi (1972) shows that young monkeys establish bonds with adult males that are almost as strong as those with adult females, and they do so despite the fact that females show a great deal of affection toward the young while adult males show very little. Until very recently the only system that commanded attention was the mother-child system. Behaviorists as well as psychoanalysts took it to be self-evident that the innate reason for the child's attachment to its mother was nothing but hunger; the child's hunger, manifested by crying, is appeased by the nurse who provides the breast. But Bowlby (1958) and

Harlow (1958) called these accepted ideas into question on the basis of direct observation, too often ignored. Both stressed the security and comfort derived from contact.

The importance of such contact was clearly revealed by Harlow's (1958) experiments with artificial mothers: a wire frame covered with furlike cloth is sufficient to calm completely the anxiety of a young rhesus monkey deprived of his mother and make him almost normal. Actually, contact alone is not sufficient; the proxy must be warm or at least permit the young monkey to warm himself upon contact with the covering. If it is cooled by circulation of a liquid refrigerant, the young monkey will embrace it only briefly, very quickly reject it, and refuse to go back to it. Then he will choose an electrically heated bare frame over the cold covered one. That is the only time when a preference is shown for a wire proxy.

9. Separation Mechanisms

At some point attachment for the mother must cease and be replaced by heterosexual and other attachments. One of the drives that leads to gradual separation is the exploratory drive, which appears in the young animal by fits and starts, as it were, interrupted by precipitous retreats to the mother. A second mechanism is the theft or "borrowing" of young animals from their mothers by other females, which is common among arboreal monkeys such as langurs. This practice is also found, though more rarely, among macaques and baboons. Lastly, a third mechanism consists of the sometimes brutal rebuffs by the mother at weaning time. Investigators differ on the relative importance of the need to explore and maternal rejection in the separation process. Suomi (1972), who has attempted to compare the two tendencies, does not find much evidence of a clearly defined period of maternal rejection. Other writers also think that the need to explore must be predominant. In any case, the key to understanding separation is that passing from one system of attachment to another does not simply consist in the disappearance of one and maturation of the other; rather, it involves the development of specific behaviors that suppress or alter the behaviors of the first system.

The importance of the affectional system among companions. According to Harlow and Harlow (1965), this system is the most important one; companions of the same age can almost completely make up for absence of the mother. What characterizes interaction at this age is play, which falls into two categories: (1) rough-and-tumble play and (2) approach and withdrawal. In the course of play one can see the gradual emergence of dominance and sexual approach. Young males indulge mainly in rough-and-tumble play; they are more aggressive and early on show signs of dominance and ritualized threat. Young females prefer approach and withdrawal; they are less aggressive in their play and soon show the passivity that will be useful to them in sexual approach.

The importance of bodily contact, which is so conspicuous in the affectional systems (quite apart from heterosexuality), is clearly at variance with the theories of Freud.

10. Conclusions

If a number of psychologists found it hard to accept the theories of Lorenz (1966), it is because a powerful movement deriving from Freud led them in the opposite direction: "The reason why the babe in arms wants to feel the mother's presence . . . is simply because he knows by experience that she satisfies all his needs without delay. . . . Love originates in satisfaction of the need for food."

Ethologists have pointed out that theories like that are based on more or less unjustified assertions and not on observation and experimentation. Hull (1952) maintained that there is only a limited number of primary drives (hunger, thirst, warmth, sex) and that all other behavior derives therefrom through learning. Lorenz (1966), however, was the first to show through his research on imprinting in birds that attachment behavior (following the mother) can develop passively, that is, without the mother giving anything in return. His discovery inevitably raised the question of whether that might not also be true in mammals and man.

Attachment and Sexuality. In classical psychoanalytic theory, the sexual behavior of an adult depends very much on the pattern of his behavior vis-à-vis his mother or father when he was a child. Adult and infantile behaviors are said to be expressions of the same libidinal force and what remains to be explained is their differences.

There happen to be several reasons, however, that lead one to believe that infantile attachment and sexual behavior are quite distinct. For one thing, these two systems of behavior are activated at different times. In addition, the classes of objects to which they are directed have nothing in common. Infantile attachment appears extremely early at the beginning of life and rapidly reaches maximum intensity; sexual behavior, on the contrary, matures very late, appearing in the immature only in a partial and nonfunctional form.

A good example of the distinction between juvenile attachment and sexual attachment can be seen in does. A young doe follows her mother and continues to do so as an adult. At the same time, the mother follows the young doe's grandmother, who herself follows the great-grandmother, so that in the end the whole herd (which includes no adult males) is following the oldest doe. There is no heterosexual attachment in this since it all takes place among females. The sudden arrival of the males at rutting time is another thing entirely.

It is true that Lorenz showed that an animal imprinted to man will choose him as a sexual partner, and hence there must be some connection between infantile attachment and sexual attachment. Various investigators have demonstrated, however, that a duck imprinted to man can nevertheless address its sexual behavior to birds of its own species later on. In fact, it has been found that in mallards the critical period for the following response is confined to the first 48 hours, but the critical period for fixation on a sexual partner (quite apart from copulation, which will only come much later) occurs from the third to the eighth week. This explains why birds fixate on one companion to follow and another for sexual relations. Finally, to go back to the five affectional systems that Harlow and Harlow found in monkeys, the very basis of their differentiation is that the maturation periods and determining factors of the various systems are not the same.

APPENDIX:
A NOTE ON IMPRINTING

The phenomenon of imprinting (*Prägung*) was discovered by Heinroth (1911) in goslings emerging from eggs under artificial incubation. Heinroth noticed that the goslings had scarcely hatched when they began to tag after him as he moved about, just as they would have followed their mother. Lorenz (1966) later made systematic observations of several species of birds and found that in the first moments of life the look of the first moving object the young birds see is indelibly imprinted on them and they proceed to follow it (a cushion pulled by a string, a man, or whatever).

His work has given rise to a multitude of studies which I cannot summarize here. I shall confine myself to research on imprinting in mammals because it is not as well known.

1. IMPRINTING IN MAMMALS

In 1963 Shipley found that guinea pigs four hours old will follow a moving white disc. In fact, they not only follow it, but also sniff it, lick it, and seek contact with it. In another experiment in which the young guinea pigs showed the same behavior, they had been kept in total darkness, which rules out any possibility of visual learning.

Scott's (1963) experiments with young dogs are less rigorous, but his findings are similar. The point in question was whether a puppy who has never seen a man will nevertheless go to one. Puppies raised in isolation from man were exposed to a seated immobile man: the younger puppies, three to five weeks of age, all approached the man; the older puppies, however, showed signs of fear. The phenomenon in question was somewhat different from classical imprinting in which a moving object is followed. Fisher (1955), though, raised young puppies in complete isolation, feeding them so that they saw neither man nor congener, and when at three weeks they were exposed to a man walking, they all followed him.

The experiment of Cairns (1966) with lambs yielded similar results. From the age of six weeks, lambs were raised in isolation except for the presence of a television set or a dog, with or without a fence between the subject and the stimulus. In every case, when the lamb was separated from the TV or the dog, it would bleat plaintively and look for the object or animal to which it had become accustomed, although obviously it could expect no reward (on the contrary, the dog would often bark when the lamb approached, and even bite it, but with no effect whatever on the lamb's attachment to the dog).[1]

2. Theoretical Conclusions

Many debates have been waged over imprinting and Lorenz's theories, and they continue still. These are some of the conclusions investigators have reached.

1. Lorenz holds that imprinting is innate and not acquired, but it seems that this may not be true in every case. In birds, for example, especially chickens, the mother's calls are undoubtedly perceived by the chick while still in the egg, which may have an influence on its subsequent behavior. Thus it is incorrect to say that in all cases behavior immediately after birth is nothing but the expression of the chromosomes.

2. According to Lorenz, imprinting is indelible, which would mean that a bird that has been imprinted to man, for example, will later be unable to accept its cogeners as sexual partners. In reality, though an animal that has received an abnormal imprint such as man is deeply disturbed in its relations with its congeners, it nevertheless can recover and

---

[1] It should be added that imprinting also exists in insects. For example, Jaisson (1974) has shown that the worker-cocoon contact of nymphs governs subsequent behavior and can result in newly hatched workers accepting (and being accepted by) species very different from their own.

achieve normal behavior. If such an animal is kept among congeners for a long time without seeing man again, it will be able to mate normally. It is true, however, that before recovery an animal imprinted to man will treat him exactly like a partner of its own species: a jay used to bring worms to Lorenz and try to stuff them in his mouth, a ritual of male and female jays at courting time.

3. Lorenz also contends that imprinting is a very special type of learning without counterpart, since it is accomplished without reinforcement and is instantaneous. That is quite incorrect. There are numerous types of learning that may be accomplished straight off and at times with no reinforcement (for example, exploratory activity and latent learning).

The fact remains that imprinting is a common phenomenon and, though one must beware of interpretive exaggerations, it is essential that one take it into consideration when studying young animals. This is especially important in view of the fact that the sensitive period during which an animal fixes on an object is extremely variable, depending on the species and sometimes even on the experimental conditions. It is often merely a matter of days or even hours.

PART III

SOCIAL LIFE

CHAPTER 7

Social Organization

Simians are basically social animals. Harlow (1958) found that merely the sight of another monkey was sufficient reward to induce learning. In nature, even sick or injured animals will make every effort to keep up with the group, and rhesus monkeys who have been unable to rejoin their own troop have been known to integrate themselves into a troop of langurs.

1. CHIMPANZEES

A chimpanzee troop may have from 4 to 14 members who no doubt make up a polygamous family, with one male dominating several others of lower rank plus females accompanied by their young. It is quite unusual to encounter an isolated animal, and such solitaries are generally old and sick, for chimpanzees are strictly social creatures.

Bands inhabit well-defined forest territories where they build nests of leaves each evening. Though baboons

A much more detailed study of social behavior in animals will be found in my book *Le comportement social chez les animaux*, published in a second edition (1973) by Presses Universitaires de France.

are carnivorous on occasion, chimpanzees are unique in hunting regularly to obtain meat. Van Lawick-Goodall (1965) has seen them gather round a male tearing up his prey and stretch out their hands for a portion. They even hunt monkeys such as the colobus.

Use of tools. Most chimpanzees throw things such as stones, sticks, or fecal matter, sometimes with very accurate aim. Gorillas do the same and orangutans as well, but to a far lesser degree. Some can even hit a moving target by aiming in front of it.

Chimpanzees use stones to crush scorpions, which are then eaten, to collect honey amid swarming wild bees, and to enlarge the entrances to ant or termite hills in order to get at the insects, a favorite tidbit. Similar behavior has also been observed in baboons, mangabeys, and colobus monkeys. Baboons will use a stick to dislodge insects hiding under rocks.

Van Lawick-Goodall has seen young chimpanzees learning to "fish" for termites with a straw, both by watching others and by experimenting on their own. Baboons often sit beside chimpanzees carrying on this operation and watch them attentively, but as fond as they are of termites, baboons never learn to do it themselves.

2. Macaques

In nature, rhesus macaques (*Macaca mulatta*) form clans of 20 to 100 animals who live mainly on the ground but also in trees in the forest or savanna. Their diet is chiefly vegetarian, supplemented by insects, birds' eggs, or small animals. Each macaque provides for itself, except for the young, who are cared for by their mothers for a very long time. There are several dominant males in each clan who observe a strict hierarchical order among themselves. The other males form celibate groups attached to the horde. Aside from copulation, the dominant males pay no attention to the females except when danger threatens. Then they will fiercely defend the group's territory. Females in estrus that mate with the ruling males become dominant themselves, but only temporarily, returning to subordinate status afterwards. Showing the hindquarters is a sign of recognition of the dominance of a ruling male, who does not necessarily take advantage of the position to copulate, but will cease an attack.

Mothers teach their young to select appropriate foods. Their diet is vegetarian but varies widely. In this connection, an interesting observation made by the Japanese is worthy of note. Within a single species (*Macaca mulatta*), one group dug up certain roots on the hillsides at Minoo; the nearby Takasakiyama group never did this, although these same roots existed in their territory; another group ate rice; and a fourth group would pass by rice fields without showing any interest (Kawamura, 1959).

a. *Acquisition of new feeding habits.* Itani (1959) studied this subject at Takasakiyama. At first, more than 50% of the young monkeys took candy that was offered them, while only one female (out of 66) and three males (out of 37) did so. A year later, 92% of the young accepted candy, 50% of the females, and 32% of the males. It should be added that among the males, those who came in closest contact with the young acquired the habit most quickly. A habit is acquired much more rapidly, however, if *the leader is the one to introduce it*. For example, at Minootani where that was the case, the habit of eating wheat was acquired in four hours! The dominant females, in close contact with the leader, are the first to become acculturated. Young monkeys come next because they roam everywhere and come in contact with everyone. Subordinate females learn the habit from their offspring.

Intention seems to have been at work on the island of Koshima where monkeys gradually learned to swim to collect sweet potatoes thrown into the sea. They first went into the water to wash potatoes left on the beach by scientists. They also learned to wash wheat strewn on the beach by scooping up handfuls of wheat and sand and throwing the lot into the water; naturally, the sand sinks to the bottom and the wheat floats on top, where the monkeys have only to skim it off. It is also interesting to note that although these monkeys are quadrupeds, those who acquired the habit of going to the sea to wash their potatoes held themselves progressively more upright as they carried food to the water. Koshima, it should be added, is an isolated, uninhabited island and no one taught anything to these monkeys.

The *environment* in which monkeys live may influence their behavior. For instance, it has been observed that young wild rhesus monkeys living on the outskirts of towns play more than they do in the forest. This may simply be due to the fact that food is more plentiful and readily available (by stealing from market stalls), while in the forest a monkey usually requires much more time to feed itself adequately. When the fruit in the forest is ripe and abundant, rhesus play a great deal. A season comes, however, when sesame seeds are almost the only food that can be found, and it takes a long time to gather these minuscule seeds; they then play very little.

Intergroup relations are also influenced by the environment. In areas where food and water are limited in supply, the monkeys are in poor physical condition and often covered with battle wounds; but in forests where living space is ample, they are much healthier and inter-group fights seem to be very rare. It should be added that the system of intergroup relations has been the subject of much controversy. Among rhesus, for example, one may well find some groups that will defend their territory ferociously, whereas other groups will pay little heed to trespassers. The reason for such differences is not known, but to a certain extent they seem to be due to the habitat.

b. *Sex life*. It seems that the old assumption that monkeys reproduce themselves annually may not be well founded. Actually, at least among the Takasakiyama macaques, which are the only ones that have been studied in depth, births normally occur in June and July, while sexual activity is most intense from November to March, ceasing entirely during the season of births. More important, how-ever, *in nature there are enormous differences among the various species of simians in terms of sexual activity*. In 500 hours of living with gorillas, Schaller (1963) witnessed only two instances of mating, less than one sees in a troop of baboons in a single morning. It is therefore no longer possible to accept the hypothesis advanced by Zuckerman (1932) that sexual appetite constitutes the sole factor of co-hesion among simians.

Sexual activity is characterized by complete promis-cuity[1] tempered by the system of dominance that prevails among baboons and macaques (and to a lesser degree among chimpanzees and langurs where the dominance sys-tem is not so rigid).

Very recent research indicates there may be *durable*

---

[1] Nevertheless, among macaques, and other simians as well perhaps, a male will not copulate with his mother (no doubt because she is the first dominant animal he ever encountered in her role as disciplinarian when he was little).

family ties within the horde and *recognition of kinship. It now seems clear that the bond between mother and child lasts throughout life*, forming a sort of nucleus out of which other types of social relationships develop. When it is possible to determine the genealogy of individual animals, one finds that very often those that stay grouped together are related (Yamada, 1963; Imanishi, 1957). Sade (1967) has seen a female rhesus defend an adult son from attack by a dominant male. Many rhesus monkeys invariably seek out related animals to groom them. We are now at a point where primatologists are beginning to wonder *whether dominance may not have been overemphasized, because of its dramatic aspects, and parental relationships too neglected*, for although family relations may take up comparatively more time, they are also much more discreet. In any event, *paternal* care, particularly among rhesus, seems to enhance greatly the subsequent status of a young monkey as compared to one that does not receive the attentions of an adult male.

3. GIBBONS

Gibbons also live in groups, but in their case the group is a monogamous family that seems to be highly cohesive and very long lasting. It consists of a male and a female along with the current offspring (there being only one birth every two or three years). Territory is well defined and defended with vocal and physical threats against encroachment by neighbors. Gibbons bed down nightly in the treetops but, unlike most apes and monkeys, they do not make nests of foliage. The vocalization and mimicry of gibbon troops are particularly rich.

4. Howler Monkeys

Much less aggressive than macaques, howler monkeys (*Alouatta*) live in groups in the tropical rain forest. Males show little evidence of dominance and do not even compete with one another in mating. They participate to some extent in the care of the young, but are never aggressive toward them as rhesus so often are. The males blaze the trail through the forest, calling out to indicate the way to the females and young who follow behind. Their most curious characteristic is the howling session at break of day, which doubtless serves to warn other clans of their presence, a fact their neighbors might not be aware of because of the denseness of the forest. Territory is marked out and guarded most jealously. It should be noted that sexual dimorphism is much less pronounced than in macaques, and howlers are correspondingly less aggressive; however, they will hasten to the defense of a fellow under attack, whereas macaques defend only their territory, not individuals.

Dominant animals can lead the whole group to manifest behavior similar to their own. One howler, observed by Carpenter (1934), used to lead the entire group in raids on all their neighbors. After Carpenter removed him, however, the leaderless group kept within the bounds of its own territory.

5. Dominance in Primate Troops

Social rank is not so rigid that small changes are not occurring continually. For example, just as young hens with chicks take precedence over older hens, female baboons in estrus or carrying their young enjoy a special status. Also common in bird or mammal is the hierarchical promotion accorded a female when she mates with a male of higher rank. Sometimes a rise in the hierarchy comes from what might be called inventive genius. Van Lawick-Goodall (1971) reports that one of the chimpanzees she observed in the wilds had come by a tin can. He would create a great racket with it, putting all his fellows to flight since they abhor loud noises. As a result, he advanced considerably in rank and later maintained his status even without his can. Not only strength but also intelligence counts in the hierarchy. Among rhesus, a monkey may gain status by forming alliances with one or more others. The technique consists in pretending to attack other monkeys while keeping close to the one being sought as an ally. (Females, however, do not form such friendships among themselves.) An inferior also tries to win the friendship of his superior by joining in when the latter attacks another animal. This sometimes ends up in very complex situations such as those observed by DeVore (1965) in baboons. Three allied males were the group leaders, but a fourth male held the highest rank *as an individual*; however, he ceded precedence to the three others when he met them together.

In monkey populations, what could be called "friendship" or "affiliation" between two males often plays an important role in establishing a new social order, because by supporting each other they can win out over the dominant male of a troop. In other situations, the relationship between brothers may be equally important. During a study that Wilson (1968) made of troops of rhesus monkeys on the island of Cayo Santiago, he noticed that the young males usually move to the periphery of the larger groups. There, where they are subject to repeated attack, they are taken under the wing of another male that has been established there for a longer time; often the older male comes from the same troop as his protégé and they may be brothers.

Some subordinates make use of a wily maneuver that might be described as a "subterfuge" to climb higher on the social ladder. They take a great deal of interest in the young of the troop and let them climb on their backs, but by and large these are *the offspring of a higher social rank than their own*; in this way the inferiors ingratiate themselves with their superiors. Among rhesus, this phenomenon of adoption is cultural, according to Itani, for it is far from being characteristic of all troops.

The terms dominant and subordinate, however, are often not particularly applicable to primate relationships, and several writers think it would be better merely to describe the "roles" played by various animals. For instance, the role played by the so-called dominant animal would be keeper of the peace, the one who stands between the troop and dangers from without and imposes peace in the event of fighting within the group. It should be added that these roles are not permanent and may well change.

In another connection, a ritual related to dominance is presentation of the newborn. A chimpanzee mother gives birth in seclusion, and when she later rejoins the troop with her newborn she holds out her hand, which is a ritual gesture of appeasement. The other animals come and touch her hand, which she offers palm up, and she then seems reassured. Gorilla mothers have a similar ritual (van Lawick-Goodall, 1965; Schenkel).

6. Social Cohesion

Within a group, dominant animals often maintain a

certain space around them, but in special situations closer
association is permitted. In any event, spacing is not haphazard but depends on age, social status, and kinship.
Regarding signals that *maintain social cohesion*, two points
must be remembered: (1) the meaning of a signal depends
very much on its context, and the same facial expression, for
example, may be either threatening or friendly depending
on the circumstances; (2) the importance of sexual bonds
was exaggerated by Zuckerman (1932). Investigators have
now discovered that several species of primates have brief
and well-defined reproductive seasons, with very little con-
tinuing contact between males and females; and yet in the
intervening period the group does not break up. It must be
admitted, however, that genital display, erection, and
handling of the penis or female genitals are very common
forms of behavior in primates, more so than in other
mammals. Lastly, a social bond whose importance has been
recognized only recently is the interest taken in the young
not only by females but by males (as among baboons). In
addition, Jay (1968) has found that in langur troops a
newborn is carried about by all the females of the group
from the first days of its life.

7. PLAY

The term play is highly ambiguous, because the
concept of play is extrapolated from human behavior. In
man, play is a type of activity that is basically opposed to
the idea of work; therefore, since animals do not work, how
does the concept of play apply to them? In our ignorance of
the true role of this activity in animals, we can only try to
sort out what circumstances tend to elicit or inhibit it. We
know, for example, that play occurs without any ulterior

motive, which is not an adequate explanation; still, the animal must be free from concern about cold, heat, food, sex, and sleep. In addition, there is unquestionably a "need to play," which shows that play is not simply an absence of something but a positive phenomenon. Furthermore, chimpanzees spend more time manipulating objects after they have eaten than when they are hungry, and is not handling things a kind of play? Loizos (1967) proposes that play be considered a "voluntary" activity in that neither privation nor reinforcement of any sort can induce it; if one

Figure 26. Ground squirrels at play. a: a lateral attack that is not in earnest, the partners trying to push each other over. b: wrestling. c: roughhousing in which the genuinely aggressive quotient seems a little higher. (After Steiner, 1970.)

tries to reward an animal for playing, that ends the game right there. Nevertheless *play can be triggered* by certain stimuli: for example, when a cat switches her tail back and forth, her kittens will immediately start playing with it. This kind of "invitation to play" is found in many animals. Cats and dogs, for instance, assume a half-crouching position, forelegs stiff and extended, eyes open wide, and

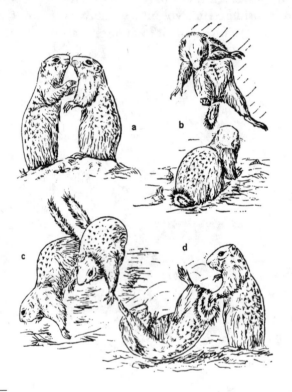

Figure 27. Ground squirrels at play. a: greeting ceremony in which two young animals investigate each other, often a prelude to play. b: playing leapfrog, c: double roll in the wake of a lateral attack, d: head over heels in a wrestling match. (After Steiner, 1970.)

Figure 28. Play sequence of a goshawk. The bird seizes an apple, throws it in the air, and, in a quick reversal, catches the apple before it falls to the ground. (After Brosset, 1973.)

ears cocked forward. A play signal will inhibit aggression. If a child pushes a friend very roughly but with a big smile, his friend's aggressive reaction will be blocked; it is exactly the same with monkeys.

Some theories hold that the function of play lies in the young animal practicing certain activities characteristic of the adult. It is pointed out that animals who have not had the opportunity to play appear to be underdeveloped as compared to others. Since adults themselves may play, however, this hypothesis does not seem sound.

Others contend that play provides an outlet for excess energy that is not used in other activities. But if energy (in the sense of physical nonexhaustion) is necessary for play to occur, energy alone is not enough, for it could just as well be spent otherwise (in exploratory activity, for example). Beach (1961) is probably right in maintaining that *no single hypothesis can explain all forms of play* in every animal species.

We can, however, examine how motor patterns are reorganized to become play activity.

1. Sequences are reordered in such a way that they do not end up in real aggression, for example.

2. Individual components are frequently exaggerated.

3. Certain components are repeated more often than the sequence usually calls for.

4. The sequence may be interrupted by movements that are not part of it and then later resumed.

5. Movements may be both exaggerated and repeated.

6. Within the sequence, certain individual movements may never be completed and such partial movements may be repeated many times over; this applies both to the beginning of an action (intentional movements) and to its end (consummatory movements).

Play is comparable to the development of ritualization in sequences where basic themes may be modified so as to become signals (Morris, 1956). According to Morris:

—the threshold is lowered;
—rhythmic repetition appears;
—certain components are exaggerated;
—certain components are omitted;
—the sequence of the components may be changed;
—the speed of performance increases or decreases;
—changes in the vigor of the movements occur.

What is unique about play is that the combination or permutation of elements is much greater than in any other activity.

8. Group Effects

Series of complex phenomena almost always occur when animals are grouped together in a more or less confined space. These are called group effects and are highly variable. I have treated them more fully elsewhere;[2] here I shall confine myself to singling out a few examples from among the vertebrates.

a. *Pigeons* (Harrisson, 1938). The male pigeon gives its young a sort of milky secretion produced in its crop. These glands will not function, however, if a young male has been kept in complete isolation. The same is true of the female's ovules, which will not reach maturity under such conditions. If, however, the birds are provided a congener (irrespective of sex) or even just a mirror, that suffices for the ovules and crop glands to develop normally.

b. *Mice.* The important research of Christian (1959) showed that when mice are confined with plentiful food and water, reproduction rapidly comes to a halt; the males develop very large adrenal glands, are highly excitable, and continue to copulate nonetheless, but the females bear no more young.

This has been attributed to the prevalence of excitement and fighting, which is said to cause not only adrenal enlargement in the males but also sterility in the females, who are too agitated to build nests and complete gestation.

But such theories tend to be simplistic. Ropartz (1967), in fact, found that it is possible to trigger typical adrenal hypertrophy *by simply exposing mice to the odor of a foreign group.* The effect occurs even when the experiment is designed so that the mice can neither hear nor see the others who are giving off the odor. The odor of castrated males, however, does not provoke adrenal development. It should be added that when castrated males are grouped together

[2] See my *Le comportement social chez les animaux*, 2nd ed. (1973), and *Précis de psychophysiologie* (1968a).

they display no aggressiveness, and their adrenal glands are no larger than those of isolated animals; this is also true for groups of anosmic males. On the other hand, castrated males may react positively to the odor of intact males and show enlargement of the adrenal glands. Therefore, they merely lack the power to emit the olfactory stimuli but are not incapable of feeling their effects.

A second more sensitive and faster test enabled Ropartz to reach a better understanding of the phenomenon. It dealt with the increase in activity that occurs when mice are exposed to the odor of an outside group. It was found that the smell of the urine of a foreign group is enough to trigger the reaction and that castrated males no longer release the crucial olfactory element in their urine. Even in castrated males, however, some trace of an effective stimulus does persist, for the increase in activity provoked by their odor was nevertheless higher than that found in the control group. Ropartz then demonstrated that in order to suppress the reaction completely, it is necessary to cauterize the glands of the sole of the foot, which are highly developed in mice. Furthermore, when the mice in a group are separated in metal sleeves (all the sleeves being side by side), their odor provokes only a very slight increase of activity, the same as that elicited by an isolated individual. For the most effective urinary factor to be released, there must be *direct* bodily contact between the mice. Therefore, the urinary factor would be a group odor and the plantar factor an individual odor.

Alone, the odor of intact males does not augment adrenal activity as much as grouping does, which of course permits aggressiveness to be freely expressed. Hence the emotion or stress induced by attacks must obviously exert some influence. Still, adrenal reaction to odor alone, with no attack involved, has been verified. Consequently, the stress theory constitutes only a partial explanation of the group effect in mice. It is clear that in castrated and anosmic mice the absence of any adrenal reaction to grouping can be attributed at least in part to their total lack of aggressiveness.

One question remains to be answered: do mice recognize the odor of their own group? We have just seen that they can distinguish the odor of a group from that of an isolated individual. Ropartz transmitted the odor of a group of mice back to them by the same means normally used to transmit the odor of another group: there was no reaction. The only possible conclusion is that there is indeed recognition of the group odor. It must be based on a mixture of different odors in varying proportions, the whole of which would constitute the urinary factor; the differing proportions would mark each group with a characteristic odor.

c. *Individual recognition and group recognition.* In different groups one finds varying degrees of individual recognition, such recognition being commonplace among mates. It even exists in fish, *Cichlidae* and *Pomacentridae* for example, where parents recognize each other individually but know their offspring only as a group. In gray geese, not only do mates know each other individually, they also know their young individually. Separate monogamous families are relatively rare in mammals, but they do exist, for example in gibbons; nevertheless, individual recognition seems to be very widespread even in the quite extensive clans which are frequent among mammals. In certain marsupials, such as the flying squirrel *Petaurus breviceps*, there is a clan odor as well as an individual odor; members of the group mark one another with a characteristic odor and will fight strangers. The exclusive family group of the prairie dog (*Cynomys*) is made up of a male, several females, the young of the preceding year, and the young of the current year; all the members know each other individually.

9. OLFACTORY STIMULI IN MAMMALIAN REPRODUCTION

It has long been known that smell plays an important role in the reproductive cycle of mammals. For example, Kelley found that though a ewe with lamb does not attract a ram, she will become attractive if her vagina is rubbed with the vaginal secretions of a ewe in heat. Only recently, however, has olfactory interaction been systematically studied in rats and mice.

a. *The Lee-Boot (1956) effect*. When female mice are in groups of four, one finds an increase in the number of spontaneous pseudo pregnancies; one can prevent this phenomenon by ablation of the olfactory bulbs or by isolating the females. Physical contact is not necessary for the phenomenon to occur. When there are some 30 females in a group, their cycles are very irregular and many mice go a long time without coming into estrus. Therefore grouping produces various disturbances in females: pseudo pregnancy in smaller groups and absence of estrus in larger groups.

b. *The Whitten (1959) effect*. Whitten found that females that have been grouped together do not mate as quickly when exposed to a male as do females that have been isolated. The effect can be countered by introducing a male in a wire cage into the female group. If a caged male is placed in the midst of 30 females, their estral cycles are much more regular.

c. *The Bruce (1961) effect*. If a female that has just mated is placed with other males of the same or a different strain, pregnancy is blocked and the female goes back into estrus within three or four days, as if copulation had never taken place. In this case all the offspring will come from a second male, as has been proved by genetic markings. As in the Lee-Boot and Whitten effects, physical contact is not necessary; it suffices merely to place the female in an empty box that formerly housed a male. The female is sensitive to the presence of a foreign male only up to a maximum of five days after the first copulation; after six days, the Bruce effect does not occur.

Males of another strain are much more effective than those of the same strain; in the first case 80% of the pregnancies are inhibited as against 30% in the second case. On the other hand, the presence of the male that covered the female initially will eliminate any reactions in her to the presence of other males. When a female is separated from her partner but not exposed to other males, pregnancy is not interrupted by putting her back with her partner. She can, then, recognize him.

Furthermore, these reactions will occur in total darkness and we have just seen that placing a female in a cage where a strange male has been kept is enough to produce the Bruce effect. One can therefore eliminate hearing and sight and regard smell as the primary factor in these phenomena. To be precise, the box must contain litter soiled by males. It is necessary to change the litter twice a day, hence the active substance is volatile and unstable.

10. APPLICATIONS TO MAN: PHYSIOLOGICAL SOCIOMETRY

a. *Social influences on steroid excretion* (Mason and Brady, 1964). Some time ago Mason (1959) found that in laboratory monkeys the excretion of urinary steriods was influenced by the presence of people in the room where they were housed. Over weekends, when few people were in the laboratory, the levels of urinary steroids dropped about 30%. Corticosteroid levels reflect various forms of situational stress, when learning avoidance behavior, for example. Actually, that is not the only endocrine response, for catechol amines also fluctuate: during avoidance conditioning, plasma corticosteroid and norepinephrine levels rose, while the plasma epinephrine level showed no appreciable change. The epinephrine level increased markedly, however, immediately before a session, when the animal realized it was going to begin; for example, when a preliminary blood sample was taken.

b. *Similar effects in human subjects.* As early as 1959, Mason observed that in small human groups accustomed to living together, the levels of 17-hydroxycorticosteroid excretion tend to cluster in a narrow range which varies from group to group. The subjects of this study were bomber crews. But in a similar study of students who volunteered to live in small groups, this tendency of steroid levels to cluster in a given group was not found in groups that included both sexes, whereas it was found in an exclusively female group.

c. *The effect of grouping and isolation on the development of tumors.* Starting from the common clinical observation that emotion, worry, and distress seem to heighten receptivity to cancer, Dechambre (1970; Dechambre and Gosse, 1971) attempted to construct an animal model of this "cancer psychosomatic" in mice. Grafted ascitic tumors were used in the test. It was found that isolated animals are more susceptible to tumor growth than grouped animals in that their tumors develop more (although their survival rate may, for other reasons, be higher than that of grouped animals).

Dechambre's principal discovery, however, was *isolation stress*; he found that in effect mice that are first grouped and then isolated undergo stress comparable to that of subjects that are first isolated and then grouped. Moreover the adrenal modifications are quite similar following these two seemingly opposite procedures. What matters is the change of environment in both cases. Until now, writers have concentrated almost exclusively on group effects (the effect of grouping on animals raised together from the beginning as compared to others isolated from the beginning). Isolation stress is just as important, but has been studied very little as yet.

d. *The McClintock effect.* McClintock (1971) studied the gradual synchronization of the menstrual cycles of young girls living in small groups in the same dormitory room. She found that in the beginning their periods occurred on random dates. Gradually the dates tended to cluster until they were almost coincidental for occupants of the same room. It is probable , though not proved, that the determining factor is olfactory (totally unconscious); it is already known that women's sense of smell is greatly affected (in terms of increased sensitivity) during menstruation. Thus there is reason to believe that perhaps the sense of smell does not play such a secondary role in man as one might think.

CHAPTER 8

Animal Communication

1. Mutual Recognition on the Family Level

a. *How do parents recognize their young?* It is well known that parents know their own offspring within a few days and will kill little strangers if one tries to introduce them into the brood. This is particularly pronounced in a familiar laboratory animal, the white mouse. Only during the first three or four days after birth can one get a mother to accept strange young; later it is impossible and the intruder is put to death almost immediately. On the other hand, if three females have dropped their young in the same cage, they combine their litters and suckle them one by one, turn and turn about. It is indeed a surprise when one lifts out a female to discover literally clusters of little mice hanging from her teats. What is most astonishing, though, is that the young grow satisfactorily under such conditions. Very distinct recognition of the young is a source of trouble for sheep breeders; when a ewe dies, other ewes will not accept the strange lamb and it must be bottle-fed. Yet to the human eye, one little mouse resembles another little mouse, line for line, and one lamb another lamb. We still do not know the nature of the releasers involved in this highly characteristic behavior—optical, olfactory, or other—but they are certainly very subtle.

In fish, the picture is not much clearer. Cichlids readily devour fish of closely related species that are the same size as their own young, but they learn—how, we do not know—to distinguish their own offspring. Noble replaced the eggs of a young pair reproducing for the first time with eggs of another species. The eggs hatched and the young were raised, but whenever the parents met fry of their own species they immediately ate them. This aberrant behavior was permanently fixed and prevented the pair from raising any subsequent offspring; they ate their young as soon as they hatched. It therefore seems, at least in certain cases, that the characteristics of the offspring are fixed indelibly in the memory of the parents. It is unfortunate that Noble's experiment has not been repeated more often with other species.

b. *How do young recognize their parents? Recognition among family members in birds.* Gulls become conditioned to their young within a very few days and are then indifferent or hostile to young from other broods. Furthermore, many birds recognize their young or their mate in the twinkling of an eye, as it were, when the most experienced observer cannot detect the slightest difference between them and other birds. We should add, however, that if one lives in daily contact with the same family of birds over a long period of time, one will learn to recognize them individually, but never with the speed and precision of the birds themselves. According to Tinbergen (1959), the stimuli involved must be very subtle as compared to the relatively simple releasers that govern innate reactions. Nevertheless, we do know something about them. Terns and gulls, for example, utilize general appearance and vocal tone. In a colony of terns, a great many birds are always in motion, flying about and crying, while others are at rest, sitting on their eggs without paying the slightest attention to all the hubbub going on around them. They react instantly, however, to the call of a mate, even if it comes from very far off and is barely audible in the surrounding din. By means of recordings, Thorpe has shown that each animal has an individualized call for help. Actually, Tinbergen has also observed very clear signs of recognition in gulls that were 20 to 30 meters away from each other and remained silent; it has even been suggested that recognition hinges on facial features, as in man. In fact, Heinroth (1911) conducted some observations that seem to show that a bird is unable to recognize its mate when the latter's head is hidden. He saw a swan in the Berlin Zoo attack his partner when she had her head under water, but he stopped immediately when she raised her head. Lorenz has observed the same thing in geese. On the other hand, Tinbergen rubbed soot on some baby gulls; though the parents seemed alarmed at the change of color, they nevertheless accepted their offspring, whom they perhaps recognized vocally. Unfortunately very few similar experiments have been described to date.

A chick's cry of distress is triggered by cold, and its calling diminishes as soon as a source of heat is provided. It will recognize its mother's call of alarm, however, only if it has been in contact with a hen during the first eight days of life. Beyond that time period, there will be no response. But chickens, unlike mice, do not recognize their young individually. Chicks' calls are more effective than their movements in attracting the attention of their mother. If chicks are placed under a transparent but soundproof bell, the mother takes no interest in them; however, a hen will anxiously search for a chick that is calling in distress from behind an opaque screen. Thus we see that in chickens sound signals are very important in relations between parents and offspring. Sight also plays a role, though a much lesser one it seems, for Brückner (1933) "disguised" a hen with bands of cloth but it did not stop her chicks from running after her.

c. *Parental recognition in fish.* Cichlid fry tag after the parent protecting them and try to follow it everywhere, even when separated by glass. An anesthetized fish will not attract them, however, unless the experimenter moves it slowly; they will flee if the movement becomes too rapid. It happens that the parent on guard duty always swims slowly, while the other one is much livelier. In this case, then, the type of movement is what matters and not the shape or precise details of coloration. The fry will just as readily follow a simple disc, but the distance they keep between themselves and the model increases as the size of the model increases; they always try to keep it in the same perspective.

d. *Mutual recognition by the young.* Fry also try to stay grouped together. If a glass tank containing other fry is lowered into the middle of a group of young fish, they will gather round the tank, particularly if their congeners there are more numerous. When they are very young, they will even follow artificial "schools of fry" made of balls of wax strung on a wire. They will do this irrespective of the color of the wax balls. This is even true of *Hemichromis* fry, which do use color in recognizing their parents, however, for then they respond only if red is evident on the fish's body or a lure.

e. *Alarm.* When a predator or some other danger threatens, gulls take to the air and give an alarm call which

Tinbergen (1959) renders onomatopoetically as "gagagaga-ga. . . ." Immediately, the young seek cover and hide while the parents attack the intruder with their claws or by dropping regurgitated food or excrement. Such alarm calls exist in all species. Certain variants will cause birds to flee, and tape recordings of such calls have even been used to combat concentrations of crows. In some species, however, on the contrary, an alarm call will provoke a concerted general attack on the enemy.

A wounded fish that has been bitten by a predator gives off a "fright substance" that triggers flight. It is specific and works only within the framework of the species that emits it (von Frisch et al.).

2. The Mechanisms of Communication

a. *Communication in vertebrates: Scent marks.* The habit that mammals have of marking their territory or certain objects with odorous secretions should not lead one to conclude that scent marks always have the same function. On the contrary, the purpose of marking and its effect on congeners seem to vary considerably from species to species.

Very often urine and excrement serve as marking substances. Surprisingly, scarcely any research has been done on the composition of odorous secretions in mammals, for example, that of a bitch in heat (except in the case of civet and then only for perfume manufacture and not specifically for behavioral studies).

b. *Location of scent marks.* Scent marks are not necessarily laid down near territorial boundaries, though that is the case with the lynx and the wildcat. Wild rabbits and many other mammals as well mark their entire territory. Rhinoceroses, too, leave scent marks scattered at random throughout their territory. It has therefore been suggested that the function of marking is to "inspire confidence" in the animal by making it "feel at home." Be that as it may, marks within the territory also happen to blaze the trails followed by such animals as rats, mice, and rhinoceroses.

c. *Self-marking and marking others.* In some cases, animals mark themselves. Female rabbits, for example, roll on the hillocks of droppings. In addition, though, a male rabbit may mark a female with urine and vice versa. One often finds the same phenomenon among rodents.

d. *Frequency of marking.* Frequency depends on various factors, for example, the *newness of the situation.* A great many mammals will deposit scent marks when introduced into a new cage of territory; similarly, they will mark a new object placed in their territory. Marking is also related to *sexuality.* To begin with, it coincides with sexual maturity in most cases. The glands used in marking usually show marked sexual dimorphism and males mark much more often than females. Also, marking generally varies with the season, but not always, and in some species it is carried on throughout the year (otters for example, which mate, however, only in early spring). In numerous instances females mark their territory much more frequently when they are in estrus. Yet in hamsters and gerbils at least, the odiferous glands are not essential for mating to take place.

e. *Marking and aggression.* It has been proved that in many species there is a *relationship between marking and dominance*: for example, a dominant rabbit will immediately proceed to leave his scent in an area where he has just driven off a rival. Marking must therefore play some

role in threat behavior, but what complicates the picture is
that marking can also be a response to threat. One of its
most important roles may be to *give warning* to congeners.
For example, Müller-Velten (1966) found that mice avoid
the odor of urine from mice under stress; similarly, a dog
will avoid the odor of urine from a frightened dog (Dono-
van).

f. *Reactions to scent marks.* One cannot always verify
the attractiveness of one sex's scent mark for the other sex. A
cat in heat, however, will show characteristic reactions
when placed in a cage that was formerly occupied by a
male; and everyone has seen how the urine of a bitch in
heat can affect a dog. Also, mice are reportedly attracted to
a trap that has previously caught a mouse of the opposite
sex.

On the other hand, it can happen that the scent of one
sex will *drive off* the other sex. On the whole, however,
scent marks probably merely indicate how long ago an animal passed the marked spot; this signal is then interpreted
in various ways depending on the circumstances and the
species. One cannot always detect signs of fear in subjects
exposed to scent marks of an animal of the same sex. When
an animal comes across a mark, its behavior does indeed
change, not to fear or flight however, but rather to *attention*. An animal in alien territory is more alert, as if it were
expecting the appearance of an enemy.

When an animal happens upon a scent, it will often
leave its own mark in the same spot (for instance, the dog
that lifts its leg at the corner of a wall). It depends on the
circumstances, however. In species that live in groups,
there are areas that are regularly marked by all the
members of the group. This practice of marking the same
spot can also be found in solitary animals such as cheetahs

or hyenas. This leads some writers to believe that these marking places may serve as a *source of information* about the number of animals, their sex, or their age. That is not impossible, for Hestermann and Mykytowycz (1968) have demonstrated that someone experienced in smelling rabbit droppings can recognize age, sex, biological cycle, and social status.

3. The Role of Facial Expression in Primates

Miller, Banks, and Ogawa (1962) were the first to demonstrate the importance of visual signals in rhesus monkeys by means of a very simple technique using two subjects. The first monkey could see a light signal which would be followed by an electric shock to *both subjects*, unless the second monkey, *who could not see the light*, pressed a lever. The only signal the second monkey could perceive was the anguished expression of his frightened companion anticipating an imminent shock; nevertheless, operant conditioning was readily established in those circumstances. In a later study Miller used the technique of positive reinforcement. The first subject could see food, but the second had to press a lever for it to be dispensed to both subjects. These monkeys were less successful, however, no doubt because the facial expression of the monkey looking at the food was not as clear-cut as the earlier monkey's expression of terror at the approach of a painful stimulus.

a. *Gaze and expression in simians.* Without going so far as to adopt the nomenclature of Kohts (1935), which is perhaps too anthropomorphic (distinguishing in simians laughing, smiling, anger, rage, surprise, disgust, terror, etc.), it must be admitted that some of these expressions are unquestionably recognizable. *Gaze is extremely important among primates.* When alert, while threatening or making a sexual approach, for example, the eyes are wide open. They are almost closed, however, in a subordinate animal under attack by a dominant one. One might say that the more self-confident an animal is, the wider it opens its eyes. As much as possible, simians avoid direct eye-to-eye contact, and when a man fixates them, they lower their heads and bare their teeth.

b. *Gaze and facial expression in man.* Are man's facial expressions cultural? Or are they innate and common to the human species? Human ethology is too young to know as yet. A menacing expression seems to be the only one that is understood as such in all cultures. In numerous studies where subjects are asked to identify the emotion expressed by someone in a photograph, the results obtained are not clear and consistent if the photo shows only the face and not the position of the arms or legs, for example. According to Secord (1958), however, subjects do fairly well in judging the character and intelligence of a person from photos.

Gaze has been studied very thoroughly. Dilation or contraction of the pupil occurs very rapidly in response to certain sights; the eye is a highly sensitive barometer of emotion that congeners know how to interpret perfectly. Hess (1965) even demonstrated that when boys are shown different photos of the same girl, their pupils dilate more if the girl's pupils have been retouched to enlarge them. And if the boys are asked which photo they prefer, they will choose the one with the enlarged pupils without really being aware of the difference. Thus very minute facial details can serve as signals and provoke emotional responses without the subject consciously perceiving the cause of his reaction.

As a child grows older he gets better and better at recognizing facial expressions. At the age of 6 or 7 years he can differentiate distress (as do monkeys), at 7 anger, at 9 or 10 fear or horror, at 11 surprise, and at 14 scorn. In an infant, the first experimental stimulus capable of triggering a smile is a card with two circles slightly separated so they resemble two eyes (Ahrens, 1954). In another connection, we should like to point out that in autistic children, who withdraw from social contacts as much as possible, what they try to avoid above all is the gaze of others; if masks are hung in the playroom of such children, they particularly avoid those with protruding or staring eyes.

4 . GESTURAL COMMUNICATION IN CHIMPANZEES

The observations of van Lawick-Goodall (1971), who lived with a troop of wild chimpanzees, show that a precise system of hierarchical relations binds them together, although the troop would seem to be very loosely structured in that one rarely sees them gather in great numbers. Chimpanzees express themselves by calls, facial expressions, and gestures.

a. *Calls*. Chimpanzees' calls are continuous and very raucous, but not clearly specific (see section c. *Primate sounds and human language*). As for *facial expressions*; there is nothing to add to what was said above; chimpanzees observe one another's faces with the greatest attention.

b. *Gestures*. Unlike their calls and facial expressions, chimpanzees' gestures are specialized. In *avoidance* behavior, which may or may not end in flight, a chimpanzee will lower his head or hide it in his arms, or he may throw his arms in the air, hide behind a tree, or very quietly slink away. Signs of *frustration* or *hesitation* appear when a chimpanzee cannot get what he wants, perhaps because a dominant animal is present or a congener does not respond as expected; then the chimpanzee will scratch his arm or armpit, yawn, groom himself, rock back and forth, or even masturbate. In extreme cases, hesitation may turn into anger. *Anger* or *aggression* is usually confined to threats, which include staring directly into the eyes (a very important and widespread signal in primates), hand gestures (particularly the one used by man to signal someone to go away), shaking the head, and a sort of soft bark. If things go further, a chimpanzee may raise his arms over his head, shake the branches above his adversary, throw sticks or stones, or stamp his feet. In cases of actual *attack*, males are usually the attackers and adult females the attacked. There are wide variations of aggressiveness. Preceded by a charge, the attack is brief; the victim may be bitten or scratched, have its hair pulled out, or be struck with the flat of the hand. In *redirection*, a chimpanzee that has been beaten by a superior or denied access to food, for example, will threaten or attack some other congener. In the ritual of *submission*, apes and monkeys present their hindquarters to a dominant animal; this gesture may be merely suggested or, on the contrary, very pronounced, with the subordinate animal bending over and leaning on its forearms. At the same time it will let out sharp cries and turn its head toward the dominant one. Or rapid and repeated bowing may accompany the cries; sometimes an animal will even stretch flat out on the ground. The inferior then bestows a kiss on the superior's face or body, less frequently on the hand; a subordinate chimpanzee may also reach out and touch the foot of a dominant one (as Indians do). To reassure a subordinate, a dominant animal will touch its body or outstretched hand; this sometimes becomes a veritable handshake. If the hindquarters have

been presented, the dominant one will touch the genital region with its hand or foot. When a threatened animal is
very agitated and howling in terror, the dominant one will
frequently pat it on the head, the body, or under the chin
(the same gesture used by a mother to reassure her child).
These caresses may continue for a minute or more until the
frightened animal quiets down. *Kissing* the hand or face,
sometimes on the mouth, is also used by chimpanzees to
provide reassurance. Young males have often been
observed to approach food but refrain from touching it
when a dominant male is present until he taps them on the
body or chin (a ritual observed in other primates as well).
Reassuring gestures also occur in situations other than those
involving aggression, for instance, when emotions are
aroused by a sudden noise or the sight of a brawl. In the
general excitement provoked by the sight of a large quantity
of food such as a pile of bananas, chimpanzees will cry out,
pat one another, embrace, etc. An adult chimpanzee that
had been attacked was even seen hugging an infant in its
arms as if to reassure itself. Also, excited or frightened
chimpanzees will often touch their penises or scrotums, or
try to hold hands with a companion.

c. *Primate sounds and human language.* A typically
human bias leads us to attribute a precise meaning to a
sound, but *that correlation seems to exist only in man* (the
case of dolphins remains an enigma). The numerous calls of
apes and monkeys bear a much closer resemblance to calls
of other animals than to human language, that is, they are
inseparable from gestural mimicry and constitute only one
element of communication. As a matter of fact, many
primatologists maintain that *a blind monkey is seriously
handicapped but a deaf monkey scarcely at all*, for sound is
not what conveys the most information. In addition, many

simian vocal signals consist of indistinct gradations corresponding to varying degress of motivation; they do not have the clear-cut nature of a robin's threat song, for example. A spectographic analysis of primate signals by Rowell and Hinde (1962) clearly shows their gradational nature. That is not without exception, however, and in *Cercopithecus* monkeys Struhsaker (1967) has found 36 very distinct nongraduated sounds.

The extreme richness and great complexity of simian communication in expressing motivation, aggression, and hierarchical order contrasts sharply with its extreme poverty when it comes to conveying information about the environment. It had been thought that a "dinner call" existed, signaling the presence of food, but van Lawick-Goodall doubts this and believes it is only the expression of a strong but diffuse emotion.

Nevertheless, in *Cercopithecus* Struhsaker finds three distinct calls corresponding to the appearance of three different predators: a snake, a bird of prey, or a land predator like a lion. The reaction of the monkeys to these three calls is highly differentiated. In the first case, they "mob" the snake as birds do an owl; in the second case, they come tumbling out of the trees and hide in the underbrush; and in the third case, on the contrary, they scramble up into the trees. But signals of such precision have not been found in other anthropoids (though they are found in a number of lower vertebrates).

Human communication is similar to simian communication in only one respect, but there the resemblance is striking: the way in which man gives vent to his emotions through his facial expressions, his gestures, or the tone of his voice. As for his unique capacity to associate meaning with sound, it is surely due to the enormous development in man of a cerebral region of association, the angular gyrus. In children an altogether different phenomenon occurs from what one finds in young apes and monkeys. Like a monkey, an infant expresses emotion by inarticulate cries or gestures, but a child's first distinct words are to describe the environment and never to express motivation or emotion.

5. Bird Song

The songs of birds contain a multitude of extremely varied sounds, but in some cases a large part of the information can be ignored. For example, if one broadcasts the song of another male to a robin, it triggers a highly aggressive response; if, however, one plays back only the high and low notes of the song at the correct intervals but nothing in between, the response will be exactly the same. Nevertheless, the rest of the information contained in the song may be utilized in other ways by birds that hear it; for individual recognition, for example. In this connection Thorpe and Hinde demonstrated that a chaffinch (*Fringilla coelebs*) can recognize a particular male by his song. In short, one must distinguish between the *message* and the *meaning*. The meaning may vary according to the context, though the message stays the same. Some birds, however, can also change the message.

a. *Call notes*. These are quite different from song proper, which is limited to males during their reproductive period and subject to many variations, particularly through learning. Call notes, on the contrary, are stereotyped, adapted to a certain predator for instance, and they are often the same or similar in closely related species. Thorpe cites as an example the alarm calls of sparrows when they sight an owl or a hawk. The calls are completely different in those two cases. In the case of the owl, the notes are typically high and very short, well adapted for locating and mobbing the enemy. In the case of the hawk, the call notes are long and low, making it very difficult for the predator to locate its quarry; sparrows that hear those notes seek shelter immediately and set up the characteristic call themselves, thus adding to the hawk's confusion.

b. *Song.* As was said earlier, song is quite another matter. In total isolation experiments, one finds, as might be expected, that call notes are not modified in the majority of cases. But certain complex songs also remain unchanged. This is the case with the song sparrow *Melospiza*; when raised by canaries, whose song is totally different, its own song nevertheless remains completely normal. Another classic example is the cuckoo, whose song in no way resembles that of its foster parents, who invariably belong to another species. Konishi (1963) has shown that experimentally induced deafness in a chicken from the first days of life does not hinder its development of a full vocal repertoire. Other birds, however, must receive specific stimulation from their parents at an early stage if their song is to reach perfection. Once their song has fully matured, though, destruction of the hearing organs has no effect (whereas such an operation before maturation fixes the song in an imperfect state). Such *maturation* is variable. In some species, the song is fully developed in the first spring, once and for all; in other species, such as the serin, each year there comes a special time for learning new elements of the song.

This innate program (though it is imperfect and must be developed through experience) may show an astonishing degree of stability. For example, in *Fringilla coelebs*, which was introduced into New Zealand in 1862 and South Africa in 1900, the song has remained the same as in Europe. By way of contrast, another bird (*Pyrrhula pyrrhula*) learns its entire song exclusively from its own father; if this bird is raised by canaries, it will sing just like a canary. Between these extremes, a whole range of variations may exist and probably will in fact be found when our knowledge of bird song becomes more complete.

Social factors are important, and listening to the song of companions is often indispensable for maturation. In this connection, some birds even differentiate dialects; that is, they will reproduce a song peculiar to one small group that differs from the song of a neighboring group though the groups may be separated by only a minimal obstacle such as a river. Living in dense forests, *Laniarius aethiopicus major* has developed an antiphonal song or duet, with very precise alternating responses, which enables these birds to tell their position in the forest at any time.

Where does imitation stop? *In nature*, it almost always stops with congeners and does not extend to other birds. This is true even of parrots and mynas; they imitate other birds and the human voice only in captivity, not in nature.

Mutual recognition is certainly of the greatest importance both between parents and between parents and young. In the enormous colonies of sea birds, for example, parents must give their young the proper food for their size and therefore it is essential that parents and chicks find and recognize each other. Also, Coulson (1968) has shown that in *Rissa tridactyla* a female that mates with the same male as the previous year will lay more eggs and raise her young more successfully than females that change mates.

Of great significance is the recent discovery (Hutchison, Stevenson, and Thorpe, 1968) of a physiological and acoustic basis for mutual recognition by mates or by parents and young. If one analyzes the special call of terns when they have found fish, one finds that it *varies considerably from one individual to the next*, but remains constant in the same individual. This no doubt explains why a tern can tell that its mate is approaching even when it cannot see it, despite the racket of the surrounding flock. So much background noise is not an obstacle per se; we have the same thing in the so-called "cocktail" phenomenon, where a person manages to follow a particular conversation despite the din of cocktail-party chatter. It might be added that a baby tern only four days old does not react at all when the cries of the colony are broadcast, but it will respond immediately to the call of its parents by turning toward the loudspeaker.

6. The Properties of Language

As in the case of the work of the Gardners and Premack (p. 64), one may well ask to what extent the modes of expression just discussed conform to the definition of language. But does such a definition exist? We can only enumerate a number of characteristic features, openly admitting that all writers are far from agreed on this list or the relative importance of its components.

a. *The audiovocal mode* characterizes human language. This method of communication has two important attributes: (1) sound production requires very little energy and (2) it leaves the rest of the body free for other activities. As we have seen, the audiovocal method is widely used by apes and monkeys, but the essence of their communication is more visual. One can follow and understand a whole series of interactions by watching a silent film showing monkeys in action; however, by merely listening to a sound band without the pictures, one would comprehend nothing. As Marler (1965) has remarked, the striking fact about monkeys is that their communication is always composite in nature, that is, it is carried on through several sensory channels simultaneously—auditory, visual, and even olfactory. Birds adhere much more strictly to the audiovocal mode.

b. *Directional diffusion and reception*. Language not only conveys a message, but this message has a destination; that is, it may be addressed to one or more specific congeners. Often vocal messages are so soft, in macaques especially, that they can be heard only by animals in the immediate vicinity. Frequently facial expression and direction of the gaze indicate the direction of the message. Jay (1968) has observed that the grimaces of langurs at play are nondirected and noncommunicative; when directed they can serve to communicate. This is also true of the yawn, which can constitute a threat among baboons; when it is not addressed to another animal, it has no communicative value. On the other hand, communication may be directed not to an individual but to a group, and Altmann (1965) finds that nowhere is this sort of group communication as highly developed as in the primates, except perhaps in certain birds. It should be noted that one of the attributes of sound signals, deep tones in particular, is that they can pass around obstacles. This is especially important for monkeys and birds of the forest that cannot see one another. Monkeys of the savannah use mainly visual signals.

c. *Interchangeability*. With language, each member of the community is both sender and receiver. This is not the case with visual releasers nor even with songs of birds as a rule (male and female have quite different songs which are not at all interchangeable). Call notes, however, or the normal cries made by members of the same group apart from mating activity, are interchangeable and serve to maintain cohesion. Still, one particular cry of the howler monkey can be produced only by males.

d. *Integral retroactivity*. This simply means that a speaker hears what he says, or perceives what he sends, whereas he does not, for example, see himself as others see him. Thus there is a big difference between sound signals and visual releasers such as the callosities on the buttocks of a female ape which only the male can see.

e. *Specialization*. This is the property of a signal being only a signal and nothing else. Among primates, some activities are specialized and others not.

f. *Semantic value*. This is the fixed correlation between elements of the message and elements of the real world. Alarm calls have semantic value. So do the initial components of a movement for, owing to the redundance of the complete movement, they can signify it in its entirety.

g. *Arbitrary designation*. There is no relationship other than an arbitrary one between the form of the message and the corresponding event in the real world; the double bark of a baboon that has seen a leopard is not the leopard. Visual releasers are just as arbitrary as human words.

h. *Digital or analog communication*. Most languages are discrete (as opposed to continuous), that is, the signals are separate and distinct and do not blend into one another by imperceptible gradations. Such communication is sometimes called a digital system as opposed to an analog system. In the dances of bees, however, there is a continuous gradation in the speed of the dance as a function of distance; thus it is an analog signal. But true analog communication would imply a continuous gradation not only of the signal but also of the response. Such communication seems to be rare except perhaps in connection with the alarm calls of baboons. Facial expressions in man and animals are analog in nature (a look of surprise that turns into one of fright, for example).

i. *Displacement* into the past or future is one of the most striking features of human language, but it is not entirely absent in animal communication. An alarm signal is sounded well before the real danger arises, and the dance of the bees refers to the discovery of nectar, an event many minutes in the past.

j. *Openness.* Our language is open-ended in the sense that we can say things that have never been said before and still be understood. One of the reasons this is possible is the fact that meaning varies as a message is incorporated in different sequences. According to Altmann (1965) and Itani (1959), Japanese macaques respond to certain series of vocalizations differently than they respond to the separate components.

k. *Duality.* This refers to man's capacity to derive new terms by combining a short meaningful element with another element (particle, prefix) having no fixed meaning. This could be said to exist in apes and monkeys. For example, if a young animal threatens an adult, the latter will not respond with a counterthreat as he would if he were threatened by an adult; in that case, then, age acts as a differential element.

1. *Prevarication.* The possibility of lying is very extensive in human language but almost totally lacking in animal communication. Nevertheless, the quail that pretends to be injured in order to distract a predator and the fox that pretends to be dead in order to catch a crow—a very common trick!—seem to be quite capable of prevarication.

m. *Reflexivity.* This is the capacity to communicate

about the communication system itself, to talk about language. Clearly, this capacity belongs exclusively to man—who both uses and abuses it.

n. *Learning.* The communicator can learn a new language. This is widespread among young animals, for example, birds who learn part of their song from their parents.

o. *Tradition.* Linguistic traditions pass from one generation to the next in man, and in birds as well.

Animals, therefore, possess some of the characteristics of human language, but man alone possesses them all. Reflexivity is the only specifically human characteristic.

7. INSECT SOCIETIES

a. *Chemical communication: Pheromones.* Insects have a means of communicating chemically through scent that is very similar to what we saw in vertebrates. Chemical communication in insects, however, seems more sophisticated; it has also been more thoroughly researched. *Pheromones* are a class of substances that excite insects chemically and govern some of their behavior. They are comparable to hormones in that in many cases their action is identical to that observed when a hormone is secreted *within* the organism, the difference being that pheromones are discharged *outside* the body by specialized glands (hence the derivation of the term from the Greek *phero*, bear or carry). Biologists, however, have included so many substances in the category of pheromones, without bothering to classify them according to function, that the term has become vague and lost much of its usefulness.

b. *Group pheromones.* German cockroaches are attracted by the pheromone emitted by their nymphs. This is also true of silkworms and the cricket *Acheta domesticus*; with the latter it seems that too strong a concentration of the pheromone causes dispersal.

c. *Attraction to the laying site.* The female mosquito discharges a pheromone into the water at the same time that she deposits her eggs, thus attracting other females to the same spot. In the migratory grasshopper *Schistocerca* and the fly *Lucilia cuprina*, the females lay en masse in the same place; a pheromone acting at short range is what attracts them.

The attraction of pheromones has been closely studied in the case of Scolytidae, small beetles that bore into the trunks of freshly felled trees. The first invaders of the tree may be either male or female, depending on the species. The initial attack is triggered by attractive substances from the tree itself, among them a number of terpenoids. But as soon as the beetles begin to feed, they discharge a pheromone through the posterior intestine, and it is this substance that unleashes the mass onslaught by other insects of the same species; males and females seem equally attracted by it. When the second wave of invaders alights on the tree, those of the same sex as the first arrivals bore their own tunnels and discharge still more of the pheromone. Those of the opposite sex seek out the galleries already carved out and enter them to mate. It is a frequent phenomenon in coleoptera for one sex to emit a pheromone that attracts both sexes.

d. *Attraction during swarming.* In bees, the formation of a swarm is stimulated by an odor from the Nassanoff glands. This odor is attractive to workers and queen alike, but workers are also attracted by an odor from the queen's mandibular gland, which enables them to stay in contact with her. Another odor, peculiar to the hive the workers and queen came from, also plays a role. If swarming workers lose contact with the queen, it can be re-established through the secretions of the Nassanoff glands and the queen's mandibular gland. From a long distance, workers are attracted by the acid 9-ceto-trans-2-decenoic. At a short distance the attractive agent is the acid 9-hydroxy-trans-2-decenoic.

e. *Orientation by reference to air-borne odors.* Anemotaxis has been definitely proved in only two insects, the fruit fly *Drosophila* and the yellow-fever mosquito *Aëdes aegypti*. In other cases it has been assumed that when insects lose a scent trail they are following, they zigzag back and forth in order to find it again. And in point of fact, when Farkas and Shorey (1972) experimentally created a scent trail in a tunnel, they saw moths (of the pink bollworm) actually perform such zigzag maneuvers as if they were trying to stay within range of the attractive odor. Wright (1963) has even suggested that these trails, which tend to fray out along the edges, provide flying insects with a directional index. The wisps of scent are denser near the odor's source than far away from it; when an insect zigzags along a trail; the number of wisps it crosses can inform it whether it is going toward or away from the source of the odor.

f. *Pheromones and dispersal.* Instead of bringing insects together, pheromones may do just the opposite, that is, disperse them. For example, the female of some species (parasite hymenoptera) produces a pheromone that dissuades other females from laying where she has already laid. In beetles of the genus *Tenebrio*, the male emits a pheromone that he deposits on the female while mating; this prevents other males from being attracted by the female pheromone. In the coleopter *Pterostichus*, the female emits a repellent pheromone which drives off males when she is not disposed to mate. This pheromone may even cause a truly toxic reaction in a male, which is perhaps an adaptive function, for the male has a strong tendency to eat the eggs and it can be important that the female keep him away until she has finished laying. The alarm pheromones of ants may cause them to gather or, on the contrary, to disperse.

g. *The role of male pheromones.* In certain cases, male pheromones do not seem to have a specific effect on that portion of the female's nervous system that controls copulation, but they influence other functions instead. For example, a pheromone may immobilize the female, thus making copulation easier. This is frequently found in butterflies, such as *Danaus gilippus berenice*. Initially the attraction is visual, with the male pursuing a flying female, but as he does so he sprays a pheromone around her which makes her land; the male then discharges more of the pheromone (from the posterior end of the abdomen) which immobilizes her completely. In cockroaches, the male everts and exposes a gland under the wing sheath which attracts the female. When she licks the secretion, it immobolizes her while the male copulates.

h. *Bee dances.* Unquestionably, the dances of bees constitute the most astonishing type of communication known among insects. In this connection I refer the reader to my discussion of the subject in *Le comportement social chez les animaux* (1973), or, if further details are desired, to *Traité de biologie de l'abeille* (1968b).

PART IV

CHAPTER 9

Ethological Concepts and
Problems of Definition

Ethology cannot escape a continual re-examination of concepts that seemed most obvious, most useful, most firmly established. This is primarily due to its methods, for in interpreting the kinds of behavior they observe, ethologists must resort to analogy and categorization—perhaps premature. The pitfalls of purely verbal explanation or circular definition are particularly great in ethology. In closing, therefore, I believe it would be useful to present a current review of certain problems of methodology and definition.

Among the subjects we shall discuss are: problems in the concepts of drive and motivation; the difficulty of interpreting conflict situations; the intricacy of determining the ontogenesis of first reactions; the impossibility of formulating a general theory of learning; questions concerning the theory of critical periods; the utility of the concept of the innate and acquired; and, finally, the debate over Lorenzian theories of aggression.

1. The Concept of Drive

Many ethologists tend to be liberal and perhaps loose in their use of the notion of *drive*: it seems obvious that hunger, thirst, and sexual appetite constitute natural classifications to which one can and should relate behavioral phenomena. One must make a distinction, however, between specific drives (e.g., hunger, thirst, sexual appetite) and a kind of general drive in terms of level of activation or arousal.

Concerning *specific drives*, one should apply the notion prudently and not assume that anything is explained by simply saying an animal is hungry or thirsty. Actually, the important thing is to analyze the variations in its behavior, which sometimes show little relation to one another.

a. An example of *questionable or low correlation* can be found by examining the so-called thirst behavior of rats. It is triggered by deprivation of water, but one can deprive a rat of water in various ways: by giving it nothing to drink; by giving it abnormally dry food; or by injecting a hypertonic saline solution in its stomach. The effects on the animal's behavior are not always the same. Furthermore, how do we measure those effects? One may adopt as a measure: the amount of water a rat takes in at the end of a period of privation; the frequency with which it presses on a Skinner bar to obtain water; or its consumption of a weak quinine solution, which is usually repellant to an animal that is not thirsty. It happens that the results of these three measures are not equivalent either. For example, while the frequency of pressing on a Skinner bar and the consumption of quinine solution both increase steadily as the period of water deprivation increases, that is not true of the consumption of water from a drinking trough. Consumption of water from a trough increases rapidly up to a point, but then the intake shows no further change after longer periods of deprivation; however, the way in which the animal drinks—continually or at distinct intervals—does change with longer deprivation.

Hinde (1970) tends to agree that in such a case the thirst drive concept is of no use to us, especially when it comes to explaining such variations. Actually, it is of some use in that it permits us to group our findings; it must be admitted that these phenomena consequent upon thirst, variable as they may be, are on the whole different from phenomena following deprivation of food. It must also be recognized, although in some respects Hinde tends to disregard physiology, that hunger lowers the level of glucose in the blood, whereas thirst affects the physiological balance in a different way. One is therefore justified in using *drives as general headings, but not as explanatory principles.*

b. And the concept of *general drive in terms of level of activation or arousal*, is it of any use? To begin with, there is interaction between drives. Hunger acts on thirst and vice versa; depending on whether an animal is hungry and thirsty or only thirsty, its behavior will differ. According to Miller, rats trained to negotiate a course to get to water will drink more when they are hungry. Fear or anxiety act on hunger and thirst; hunger and thirst affect avoidance behavior; fear affects sexual behavior, and sexual factors will even influence hunger and thirst. For example, female rats trained to run through a maze in order to eat or drink will do it more quickly when they are in estrus. A convenient explanation would be that all drives play a role in an animal's general state of excitation, either raising it or lowering it. In this connection, measurement indices have even been sought, such as heartbeat and skin sensitivity. These various indices show a rather low degree of correlation. We know, however, that behavior is not normal when sensory excitation falls below certain limits, or, in other words, when the reticular system is too inactive. That observation tends to be compatible with the hypothesis of general arousal of the organism. It must be remembered, however, that the various stimuli do not contribute equally to the general excitation; it is not a matter of simple addition. Furthermore, there is no clear and consistent correlation between the general state of excitation or arousal and any and all types of behavior.

In short, the concept of *general drive may be useful in a first approach, but it must be accurately defined* and its role in each particular case made clear.

2. THE CONCEPT OF MOTIVATION

a. *How do motivational factors affect reactivity?* Hungry animals have a greater tendency to go to places where they can find food. The question arises whether this is due to increased activity provoked by hunger, in which case there would be a greater chance of the animal coming across places where food is to be found. Or is it perhaps that hunger causes greater susceptibility to stimulation, which would in turn result in increased activity?

Campbell and Sheffield (1953) question the first hypothesis, whereby hunger increases activity. Hungry rats move about no more than control rats if placed in a totally *monotonous environment*, but, if the *environment is varied*, much more than the control animals. Although this result has been confirmed by several investigators, using monkeys in particular, it must nevertheless be added that even in a milieu very poor in stimuli, hunger still produces a slight increase in activity. It is just that the increase is *much less pronounced* than in a varied environment.

b. It seems that what changes most as a function of motivation is *the response to certain stimuli* to which animals become highly receptive. In this connection, Baerends (1957) observed a very curious phenomenon in cichlids. During the parental phase they may stand guard over daphnia or try to lead them about, just as if they were their fry, though normally cichlids feed on those water fleas.

c. It had been thought that the *sensitivity of sense organs* changed with motivational state. In man, however, Meyer (1952) found that after 34 hours of fasting there is no change in the threshold of sensitivity to salt, sweet, or bitter. In another connection, removal of a rat's gonads does not alter its ability to distinguish between a female that is in estrus and one that is not; in a maze the castrated rat will still seek out the receptive female. What is affected, then, is the *response to a stimulus* rather than the *sensitivity to stimuli* per se. This implies that the stronger the motivation, the sooner stimuli become adequate that were not so before.

3. Conflict Situations

It frequently happens that an animal finds itself in conflict between two tendencies, for example, the tendencies to flee and to fight when it encounters a rival at the frontier of its territory. But it is not always easy to determine if there is a conflict or not. First, one must consider the situation: combat on a territorial frontier between two males in the presence of a female, etc. Then behavior must be taken into account: if, for example, a subject alternately approaches and retreats from an object, there must be a conflict. In certain fish, one can sometimes utilize the changes of color as a barometer of emotion. Behavior before or after a certain action or display can also provide clues to conflict; for example, if a display is followed sometimes by attack and at other times by flight, it obviously expresses a conflict between two tendencies. One can even get an idea of the strength of the opposing tendencies by counting how many times they are each expressed within a given period of time. Finally, *the nature of a display* should be considered. A display generally consists of a sequence of movements or postures, but each component may be more or less complete; for example, the lift of a wing may be barely hinted at or, on the contrary, very pronounced. Below are some of the various possible ways in which conflict is manifested in display.

a. Certain components are actually elements that can be borrowed from differing situations. The gull's tensed posture which characterizes threat may be accompanied by raised carpal joints, neck extended forward and up, bill pointing down: these constitute attack components. On the other hand, if the head is oriented more laterally, the feathers flattened, the bill tilted up and the neck stretched up as well (instead of a more forward thrust), such behavior corresponds to flight. The fact is that threat posture contains elements of both behaviors, and one can see how easily one or the other may predominate.

b. If as a rule a certain component is never seen in display *a* but always in display *b*, and yet nevertheless it is occasionally observed in *a*, one may conclude that normally it does not belong to display *a*.

c. The particular behavioral situation in which a component appears most clearly in association with a certain activity, such as fighting or copulation, permits one to associate it by and large with the behavior in which it is most conspicuous.

d. Finally, comparative study of different species will provide the best information on the role and meaning of a display or its various components.

Wiepkema (1961) attempted to clarify these rather vague notions by *factorial analysis* of the reproductive behavior of swordtails (*Xiphophorus*). He identified 12 different movements and found that only three factors accounted for 90% of the variance. Four movements (butting, pursuit, abrupt turns, oscillation) are grouped around factor 1, and since they are manifestly aggressive one can say that factor 1 is linked to aggressiveness. Four other movements not related to reproduction, flight in particular, are grouped around factor 2. Finally, the last four movements, which are found in courtship behavior, are related to factor 3, which can therefore be called the sexual factor.

It has been noted that a great many social postures seem to show signs of conflict. In particular, one frequently observes competing tendencies toward fight and flight. Nevertheless, one should not connect all social situations with conflict; certain activities, such as grooming or laying down scent trails cannot be put in the context of conflict. From the neurophysiological point of view, however, we must also mention the very interesting experiments of Brown et al. involving direct stimulation of the central nervous system of cats. *Attack and flight are triggered in adjacent regions.* As stimulation is moved to the posterior, the response includes more flight components and becomes less aggressive. Therefore these differing behaviors are closely related physiologically.

As for the old idea that conflict is due to incompatibility between two opposing reactions, such as attack and flight, it is better to say that certain responses of these two tendencies (but not necessarily all) are incompatible.

4. Conflict Behavior

a. A great many different types of behavior can be observed in conflict situations. For example, *inhibition of all responses except one*: when sparrows that are feeding see a predator, they abandon everything to seek cover. A very strong conflict can result in inhibiting all responses, as in the paralysis caused by intense fear. One also observes *intentional movements*: a bird in a conflict situation may make the gesture of taking flight without carrying through. *Alternation* between two movements is evident in many instances. This can be seen in a rat that is first trained to go through a maze to obtain food, and then, once that is mastered, is given an electric shock when nearly to the goal; afterwards, when the rat reaches the middle of the maze, it will "oscillate" between going ahead and turning back.

b. *Ambivalent behavior* is seen when, instead of alternating, the animal performs different or even opposing movements at the same time. For example, if wild moor hens (*Gallinula chloropus*) are given grain, they will begin by approaching it, then pecking it or even swallowing it, and at the same time start to flee. *Compromise behavior* expresses in a single gesture what ambivalent behavior expresses in several. An example of this can be seen in the phototropism displayed when an animal is exposed to two light sources: it follows the bisector of the angle formed by the two sources. In other cases, conflict between approaching a goal and avoiding it may result in *circular movements* around the goal.

c. *Displacement activities* are observed when animals in a conflict situation perform movements "out of context." For example, sparrows in conflict (aggressive behavior, sexual behavior) may rub their bills against something hard or preen their feathers. It is not easy to interpret such behavior; three explanations have been suggested.

1. One hypothesis holds that displacement behavior is the consequence of some *autonomic activity* being triggered by the situation. For example, vasomotor changes in the skin may stimulate the skin, which would induce scratching or grooming.

2. According to the hypothesis of *disinhibition*, when conflict blocks the appearance of types of behavior that would have first priority, other activities that are normally blocked may be disinhibited and appear. This hypothesis was carefully investigated by Sevenster (1961) in connection with the male stickleback fanning the nest. Fanning movements are normally meant to irrigate the eggs, but they also occur well before the female has deposited her eggs in the nest and during courtship as well. Sevenster attempted to measure the strength of the different tendencies. He first measured the amount of fanning, which he took as an index of the parental drive (P). He also evaluated the sexual drive by exposing the male to a female and counting the number of zigzag maneuvers he made (S). He found that at low levels of S, S is in direct correlation with P; in other words, displacement fanning depends on the same factors that influence parental fanning. But courtship fanning is in inverse correlation with the number of zigzags; this indicates that sexual drive inhibits parental activities. Aggres-

siveness also inhibits fanning when it is associated with sexual drive.

3. The third hypothesis suggests that though conflict indeed provokes the appearance of movements "out of context," it does so in a nonspecific way, by activating general drive or increasing alertness, as it were. But the results of experiments testing this hypothesis are contradictory.

5. GOAL-DIRECTED BEHAVIOR

The idea that behavior may be goal directed has become more respectable and less suspect of mentalism in view of the fact that a computer program can be designed to include an order to win at chess, which in a way constitutes a goal. It is not always easy, however, to tell whether an animal is directing its behavior toward a goal, that is, whether it connects a mental image with what it is doing. Thorpe speaks of goal direction in the nest building of long-tailed tits (*Aegithalos caudatus*) for, as Thorpe points out, it is clear that as the bird builds, its behavior is influenced by stimuli from the nest. It has also been established that irregularities and holes have a marked effect on its building behavior. Thus it becomes extremely difficult to deny that the bird must be comparing its work to a mental image (see Chapter 2).

6. THE ORIENTATION AND PERFECTION OF EARLY ACTIVITIES

Very often an animal must learn to orient and perfect an activity that is otherwise inscribed in its hereditary program. The observations of Eibl-Eibesfeldt (1970) on squirrels provide a good example. Inexperienced squirrels have a very marked tendency to hide nuts by digging, depositing a nut, pushing it with the muzzle, and covering it over—even if the whole procedure takes place on a tile floor and all these motions are in vain! With a little practice, though, things improve, and the squirrel becomes capable of hiding a nut quite successfully, even in such circumstances. Similar improvement occurs in cracking nuts. In order to open a nut, a squirrel must bite one end while twisting it around its axis. At first, a young squirrel will bite a nut more or less at random and only gradually will it develop the technique of biting the end of the nut.

In this connection, however, one must be wary of the role of *maturation*. Pecking undergoes very marked development in chickens. Although the pecking movement is in their program, in the beginning only 15% of their pecks will result in actually picking up a kernel and eating it. If one does not allow chicks to peck before a certain age, however, by keeping them in darkness and feeding them by hand for one to five days, one finds that accuracy of pecking nevertheless improves with age. Thus it is a matter of maturation rather than learning.

7. THE ORIGIN AND DEVELOPMENT OF FIRST REACTIONS

It must be remembered that a great many reactions are competely innate. For example, a sparrow that is 30 days old and has never seen an owl will manifest the mobbing response the first time it is exposed to one. The appearance of such responses is not confined to a very young age. For example, female robins, whose breasts are grey, are dominated by other females whose breast feathers have been dyed red. This phenomenon in which dominance provokes an avoidance response is also evident in females that have been fed by hand since birth and have never seen a male; the reaction cannot, therefore, be the result of social experience.

Schneirla (1965) has suggested that many of the reactions of young organisms are quantitative rather than qualitative responses. Young animals would tend to move toward sources of stimulation that are weak and constant and away from those that are strong and irregular. Consequently, as a stimulus steadily increased, it would trigger first approach and then retreat. Qualitative responses would not enter the picture until later. For example, Collias and Joos (1953) found that newly hatched chicks are attracted by a rather wide range of repetitive sounds as well as by moving objects or pictures. But they flee (or sound a distress call) when exposed to intense stimuli. Many amphibians such as toads will flick out their tongues at a small object but will retreat in the face of a large object. Schneirla's theories merit consideration; however, in many of the examples he cites there are qualitative differences between the stimuli that seem no less important than the quantitative ones.

In any case, the most striking phenomenon in the development of young animals is the narrowing of the range of stimuli to which they will respond, a narrowing in which learning—remarkably rapid at times—plays a major role. For example, newborn chicks will peck at any spot whatever that contrasts with the ground, but they soon learn to confine their pecking to food. At first, young blackbirds stretch open their bills whenever they hear any kind of auditory stimulus, but before long they will react only to the call notes of their mother (Messmer and Messmer, 1956). Young guillemots also learn their parents' call in the very first days of life.

8. THEORIES OF LEARNING

During the thirties and forties, investigators all believed, not without some optimism, that it was possible to construct a general theory of behavior. The sweeping assertions of Hull and Skinner seem astonishing to us today, but, to tell the truth, doubts were raised early on about such grandiose syntheses. Already in 1954, Koch, challenging Hull's mathematization of behavior, wrote: "In the present state of our ignorance no one can seriously believe that a comprehensive, quantitative, and hypothetico-deductive theory of behavior is possible." Logan (1959) added that most of the problems of the theory arise when one departs from the simplified situations on which the theory was based. In this connection, biologists, confronted with the enormous diversity of animals, have always been more reserved than psychologists who confine themselves to examining a limited number of situations that are invariably the same. That is why the former have strong reservations about such statements of Hull's as: "All behavior of members of the same species, and that of all species of mammals, including man, obeys the same set of primary laws."

Hinde, while refraining from laying down new laws, contributes the following observations.

1. We know that what an animal does may be continued or interrupted or modified by changes of internal or external stimuli, for example, by endocrine or neural changes or even by completion of the act in progress; the relative importance of these different factors varies with the type of behavior in question.

2. The same category of behavior may be controlled by very different factors, for example, the feeding behavior of blowflies or white rats.

3. When the learning theorists define learning as a change of behavior brought about by practice, they forget *unapparent learning* which does not affect behavior directly. One example is habituation; subsequent learning is altered by prior habituation. Depending on whether or not an animal has become habituated to a new environment, learning will be very different; if one does not take into account the exploration that precedes habituation, one risks making serious errors. A second form of such learning is *imprinting*, which occurs in relation not only to parents but also to the native environment. A third example is found in the song of adult birds that will later be imitated by their offspring. Young birds actually learn this song long before they are able to sing it themselves.

One could multiply such examples of unapparent learning many times over, for instance, latent learning,[1] the importance of which has only recently been recognized. We should add that conditioning theorists are quite aware of such learning, but they tend to relegate it to a very subordinate position. Biologists, however, believe its role to be of prime importance.

9. Sensitive or Critical Periods

Sensitive or critical periods are still widely used concepts. As defined by Scott (1963), they are periods in the life of an organism when various aspects of behavior are par-

---

[1] Latent learning is evidenced by the fact that an animal can come to learn a maze without punishment or reward. If the animal is allowed to live in the maze and feed at will, one finds that later it will be able to find its way about the maze. According to the learning theorists, though, there is no basis for its success, since there was no reinforcement (punishment or reward) for one activity or another.

ticularly susceptible to modification. Sensitive periods are not necessarily the same for all behavioral characteristics, nor are they peculiar to young animals alone. For example, in goats that have just dropped, there occurs a brief period in which the mother learns the characteristics of her newborn kid; this learning is very rapid and indelible, and subsequently any strange kid will be rejected.

In Hutt's opinion, the definition of a critical period is circular. Piaget (1936) maintains that what a child can learn at a certain stage is limited by the cognitive structures that exist at that time. But by what criteria does one define cognitive structures except by what can be learned? Actually, such criticism is unfair, at least in part, for a correlation can be established between physical *age*, the corresponding development of the *nervous system*, and learning capacities. There is at least one way, then, in which it is possible to characterize learning capability independent of learning as such. Hutt himself agrees that since these stages of maturation of the nervous system are well known in children, they can be useful to psychologists studying the ontogenesis of learning.

Nevertheless, there is nothing absolute about the definition of a critical period. When it occurs it is dependent upon the rearing conditions, for the stimuli to which an animal is exposed at a young age have a powerful influence on the biochemistry and development of the nervous system. What a child can learn at a given age depends on the way in which he is taught. By using a "talking typewriter" (which sounds a letter when the corresponding key is pressed), investigators have succeeded in teaching 3-year-olds to read. On the other hand, it has been found that a child's attention span is only 7 minutes at the age of 2, but climbs to 13.6 minutes at the age of 5. It would seem pointless to teach a child to read before his attention span exceeds 10 minutes. More recently, however, it has been proved that with highly stimulating and varied objects the limit of attention can be stretched to 25 or even 35 minutes. Thus when speaking of critical periods, one should never forget the conditions of stimulation and rearing.

10. THE INNATE AND THE ACQUIRED

For some time very heated and often quite confusing debates have been going on in ethological circles about the concepts of the innate and the acquired. The objectivists, along with Lorenz, and in reaction to the American environmentalists, tended to stress the innate character of many behaviors, "programmed" from birth, as it were. They had no trouble in supporting their arguments with a multitude of phenomena that the environmentalists had ignored because of their dogmatic insistence on the all-importance of the environment. Currently, a new offensive by the environmentalists is underway and they, in turn, are having no difficulty in criticizing the excessive rigidity of innatist theories. Here, as everywhere, truth and common sense no doubt lie somewhere in between, though the opponents are often too partisan to admit it.

a. *Isolation*. The favorite argument of hereditarianists is based on experiments involving isolation from birth. If under such conditions characteristic forms of behavior nevertheless appear, then it is a good bet that they are innate (programmed in the genome) and not acquired.

But the *validity of isolation experiments* has been criticized by various writers such as Schneirla and Lehrman (1953), who maintain that all sorts of influences may act on embryonic organisms, for example, the chick in the egg. And in fact the work of Gottlieb (1968) clearly indicates that conditioning to various outside influences may be possible before hatching. An example, which happens to be a particularly poor choice but nevertheless is often cited, is the observation of Kuo (1932) that while still in the egg a chick's bill is raised and lowered rhythmically by the beating of its heart; Kuo thought that perhaps that predisposed the chick to future pecking movements. Even authors like Hinde (1970) consider that a likely hypothesis. However, many other birds, whose bills also move rhythmically with the heart, do not peck. Furthermore, Hamburger (1963) and long before him Preyer (1885) have shown that the connection between sensory and motor neurons has not yet been completed in the medulla when the head is being moved by the action of the heart.

In any event, it is a well-known fact that ducklings incubated by a hen run to the water as soon as they hatch, while chicks incubated by a duck refuse to follow their adoptive mother into the pond; this clearly indicates that there are very precise hereditary components in behavior. From among thousands of possible examples of innate behavior, we shall cite only a few in animals of widely differing species (Eibl-Eibesfeldt, 1970). Carmichael (1926-1927) raised tadpoles under narcosis until nonanesthetized control tadpoles could swim; he found that upon coming out of narcosis the test subjects could swim as well as the controls. Grohmann (1939) raised pigeons in cages so small that they could not move their wings; he released them when control pigeons were able to fly and found that the test subjects could fly just as well. As squirrel raised by hand by Eibl-Eibesfeldt was fed a liquid diet and had never seen a nut, yet it behaved almost normally when given some. It first ate a few, then, once it was sated, it carried several nuts around in its mouth as if looking for something. It was attracted by vertical structures, corners where the walls intersected. There it would try to bury a nut by going through the typical squirrel motions even though it had never seen such motions. A hand-raised shrike removed the stingers from Hymenoptera before eating them the first time they were offered, although the bird had had no model to follow (Gwinner, 1961).

b. *Types of releasers*. Using Skinnerian methods, Sackett (1970) isolated four male and four female rhesus monkey from birth to 9 months. From time to time he projected slides showing monkeys or neutral objects (landscapes, geometric figures, etc.), and by pressing a button the monkeys could have the pictures they liked best projected again. They preferred pictures that showed their congeners, especially those of a young monkey and a threatening adult male. These pictures caused all sorts of social responses in the spectators; vocalizations, invitations to play, manual exploration, etc. At the age of 2½ months, however, the young monkeys suddenly showed reactions of fear at the sight of the menacing adult and no longer pushed on the lever that corresponded to that picture. Yet *they had never seen a threatening male*, not even their own reflection in a mirror. Surely this must be a case of an innate mechanism that becomes functional as a result of maturation (see Chapter 6).

How compelling innate mechanisms are can be seen in the behavior of turkeys. Sensitive only to the call of the newborn, a turkey hen will go so far as to sit on a stuffed ferret if it is wired to broadcast a chick's cries. A mother turkey deprived of hearing, however, will kill even her own young (Schleidt et al., 1960).

11. Specific Learning Capacities

In a study on learning, Hinde and Stevenson-Hinde (1973) investigated how innate peculiarities influence the acquisition of new habits. Their effect is sometimes more subtle than hereditarianists imagined.

a. *Animals cannot learn just anything, at any time, in any way.* If an egg rolls out of the nest, a gull will bend its head and retrieve the egg with its bill. Its bill being straight and the egg round, this maneuver is awkward and inefficient, but the gull never learns to retrieve an egg with its wing or foot. In the same vein, gulls' eggs show wide individual variations in color, but gulls never learn to choose their own eggs when presented side by side with a neighbor's eggs. Yet these same species show great accuracy when it comes to recognizing their mate or offspring.

b. *Preparedness to learn.* Depending on their species, animals seem physiologically prepared to learn some things more easily than others. Razran (1961) gives several examples of this particular type of disposition. A hare learns to run very fast after a moving object when a tactile stimulus on the nape of the neck is associated with a sound resembling the unusual noise hares make by smacking their lips. This association shows few signs of fading even after more than a year. On the other hand, if the sound of a metronome is used instead of the smacking noise, the association is acquired only with great difficulty and is very easily lost. To cite a similar case, turkey hens must hear the special gobbling of little turkeys before they will care for their young. A mother turkey rapidly becomes conditioned to this noise and will continue to care for her chicks even when they no longer make it. There are innumerable examples of this sort. Seligman (1970) suggests that separate studies be made of associations that are "prepared" (by evolution) and those "not prepared" (invented by the experimenter and unrelated to the animal's life). He thinks that laws of learning, if they exist, would apply only to the latter.

c. *What do different species learn in similar situations?* *Recognition of eggs.* As we have seen, gulls almost never succeed in recognizing their own eggs. Guillemots, on the contrary, manage to do so with no difficulty.

Ease of imprinting. Imprinting is by no means universal. Though goslings hatched in an incubator readily follow the experimenter as soon as they can move about, this tendency is much less strong in other species.

Song development. All sorts of possibilities are manifested in birds. In some species, even a completely isolated bird will produce a song without any prior imitation. In other species, however, a bird must hear a certain song at a certain time in order to sing it himself. And then there are birds (for example a myna) that are fully prepared to imitate any sort of song.

Tool using. Certain birds utilize sticks (*Cactospiza*) or pebbles (Egyptian vulture), whereas other closely related species do not. This is probably not a matter of special preparedness to learn, but rather, according to Eibl-Eibesfeldt, a question of species-specific movements used in a particular way.

Orientation in fish. Cichlids from equatorial waters (*Aequidens portalengrensis*) can easily orient themselves to the sun in the northern or southern hemisphere, though its course looks different. When raised under artificial light, certain *Centrachidae* which are found only in the northern hemisphere can orient themselves toward the sun the first time they see it, provided it is moving from left to right and not inversely as in the southern hemisphere. This does not bother *Aequidens* subjected to the same test.

Orientation in bees. According to Lauer and Lindauer (1973), two European strains of bees show very similar learning curves when food is placed in an open field, but the curves become markedly different if the food happens to be near a tree or some other easily recognized landmark. *The difference remains even when the two strains are combined in the same hive.* It must therefore be genetic.

Human language. The experiments of Premack and the Gardners notwithstanding, there is obviously an enormous difference between the language-learning capacities of man and chimpanzee.

     d. *Relative ease of association between stimuli.* At least one case is known in which an animal perceives a stimulus but is unable to use it in learning: an octopus perceives the weight of objects, but is unable to learn to differentiate them by weight alone. Monkeys learn to differentiate objects much more readily if they can not only see but also touch them. Bees learn to recognize odors most easily, then colors, then shapes. Certain birds learn the song of the parents that feed them, even if these parents are of a different species and despite the fact that they also hear the song of their own species.

     e. *Difficulty of certain responses.* Pigeons can supposedly be conditioned to anything whatever (including any animal, according to Skinnerians). Nevertheless it must be pointed out that *it is extremely difficult to condition certain responses.* Hogan (1965) succeeded in training pigeons to preen their feathers to obtain food, but the preening was incomplete and interrupted by extraneous movements. Konorski tried to condition dogs to yawn; they would open their mouths, but not really yawn.

There are differences that *depend on sex*; a stimulus that reinforces learning in one sex does not necessarily reinforce it in the other. Differences also depend on the species. In chaffinches (*Fringilla coelebs*), the song of the species is a positive reinforcement, but a certain call note is a negative reinforcement. Rats easily learn to run or jump to avoid an electric shock, but it is more difficult if they have to press a lever.

Chaffinches. Stevenson-Hinde designed the following tasks for male chaffinches:

1. pecking a key to obtain food;
2. sitting on a particular perch to hear a recording of an adult bird's song;
3. pecking a key to hear that recording.

Exercises 1 and 2 were easy. Exercise 3 proved to be almost impossible and the birds would peck on the key only if a seed was stuck to it. Therefore song does not constitute reinforcement for pecking on a key, whereas food does.

Sticklebacks. Sevenster trained sticklebacks to swim through a ring or bite on a rod. As reinforcement the fish were shown either a male ready for combat or a female ready for courtship. In these circumstances the sight of the male was adequate reinforcement for the fish to learn both exercises, but the sight of the female was adequate only for passing through the ring and not for biting the rod.

It is extremely difficult to predict in advance what will or will not serve as reinforcement with a view to applying some rule or other. All attempts to date to find such a rule show as many exceptions as cases of applicability. For the time being we must confine ourselves to compiling an inventory of effective or ineffective reinforcers. This course is only good sense. It is unfortunate that so much time and effort had to be expended before arriving at it.

f. *Facility of certain responses.* In nature, some forms of learning are almost inevitable because the various phases of a behavior are governed by the same factors or hormones. For example, nest-building behavior in birds: it must be learned, but learning is bound to take place because the various phases have a common dependence on the same hormones and also, no doubt, because the completion of certain phases is itself a reinforcement.

12. THE THEORY OF AGGRESSION

a. Controversy is currently raging in the field of ethology about the nature of aggression. Is aggression or aggressiveness innate or acquired? Some maintain that one can develop (or inhibit) aggressive tendencies at will through

Figure 29. Fright posture of a ground squirrel. (After Steiner, 1970.)

Figure 30. Aggressive behavior of ground squirrels. Top: mild interaction, limited to a light blow on the back of the adversary's head. Bottom: confrontation between two adult males in the spring, the animal on the left in a defense posture, the attacker on the right poised to leap. (After Steiner, 1970.)

various procedures or environmental influences. For example, Scott (1966) showed that aggressiveness could be greatly increased in male mice by arranging for them to win fights several times running; on the other hand, very placid males were obtained if they were housed with females and were lifted up by the tail from time to time and tapped gently with a finger. This is also true of puppies that are frequently picked up; they become very docile. Scott came to the conclusion that aggressive behavior is acquired and it emerges when animals experience pain or frustration (that is, any circumstance that prevents a behavior from achieving its goal). Berkowitz and Montague (in Berkowitz,

Figure 31. A dominant male ground squirrel perched atop a hillock in the typical "sentinel" position. (After Steiner, 1970.)

1963) concur with this hypothesis and have drawn certain inferences from it in regard to child education. *Aggression would therefore be a form of reactive behavior.*

b. Opposing that hypothesis are the theories of Freud and Lorenz, according to which *aggression is a dynamic instinct.* Lorenz in particular emphasizes not only the instinctive basis of aggression but also its role in preservation of the species.

A number of experiments have shown that rats or mice raised in isolation are very aggressive and will attack a congener as soon as one is introduced into their cage. Later ex-

periments have demonstrated that lack of play with companions of the same age is what makes these mice aggressive. In 1964 Lagerspetz discovered a genetic factor; mice
from an aggressive strain raised by an unaggressive mother
are markedly more aggressive than young from an
unaggressive strain. Therefore the influence of the mother is
not the determining factor. Jungle cocks and Siamese fighting fish (*Betta splendens*) are also very aggressive when
raised in isolation.

By no means, of course, does this rule out the possibility that social experience may affect the development of
behavior. For example, a rhesus monkey raised exclusively
with its mother is more aggressive than monkeys that have
had a chance to play with companions of the same age. And
obviously many other factors also influence aggression, such
as the fact of being in one's own territory or not, hormonal
conditions, or intracerebral excitation.

c. *Arguments against the theory of innate aggression.*
These arguments are often more philosophic than scientific

Figure 32. Postures of the gray swan. An attack by a male, the last two
figures showing the end of the attack.

in nature. For example, Berkowitz (1963) has written: "An innate aggressive drive cannot be abolished by social reforms or by removing frustration. Neither complete permissiveness on the part of parents nor fulfillment of all desires will completely eliminate interpersonal conflicts ... The lesson to be drawn from this for social order is clear: civilization and morality will ultimately have to be based on force and no longer on love and charity."

These conclusions seem extreme. Even animals that live in groups manage to neutralize part of their aggressiveness. In many instances individual recognition blocks aggression completely. According to Schenkel, lions, for example, will readily kill lions belonging to another pride, but absolutely never one of their own. That is also true of primitive man.

d. *Group cohesion.* As a matter of fact, mechanisms of group cohesion, which by their very nature oppose aggres-

Figure 33. "Triumphal" postures, after winning a fight for example, in four different species of geese. (After Veselovsky.)

sion, are numerous and well known. For one thing, females generally are not attacked, nor young animals either, and other members of the social group are attacked only rarely if at all. Also, young animals even of very different species will not attack one another if they have been raised together.

A classic example of inhibition of aggression is *hierarchical organization*. In the same category are *gestures of appeasement*, which are best known between male and female but also occur between members of the same sex. For example, grooming is one of the principal rituals of harmony among monkeys.

Furthermore, in many species the dominant animal plays a very effective role in maintaining order and reducing conflict (for example, in sea lions and baboons). The young also constitute a center of interest that binds the adults together. It has been observed in *Propithecus* that after a birth occurs, grooming becomes four times more frequent among those monkeys than before (Jolly, 1966). In addition, all animals actively seek contact with their congeners, even to the extent of direct bodily contact which is particularly apparent in apes and monkeys.

Finally, Thorpe (1975) points out that a distinction must be made between aggressive behavior (which seems to be universal) and violent behavior, which is much less common in animals than it is in man.

Summary and Conclusions

The behavior of computers constitutes the only possible analogy with animal behavior. Computers can alter the program if need be, that is, arrive at a goal by different means, and animals can do the same (as in the case of the rat that learned a maze on foot and could later negotiate it without error while swimming, although the mode of locomotion is totally different). Hence *minutely detailed observation of the actions of animals is perhaps not the sole nor even the most important task of ethology*. It will probably prove more profitable to explore the ultimate possibilities of the organic machine, not shrinking from tackling complicated problems which are more informative than simple ones (Chapter 1).

Construction is one such problem. In caddis flies repairing their nests, we can clearly see that all sorts of methods are used, varying with the individual and even in the same individual. Bees make extremely complicated repairs, finding the optimum solution by means of a complex and variable process in which numerous individuals cooperate in a way that has yet to be clarified. The same is true of ants. Birds offer us constructions of astonishing complexity, such as those of the bowerbirds, but experimental study of them has scarcely been broached (Chapter 2).

Animal learning goes far beyond what could be apprehended by the simple and naive techniques of the early in-

vestigators. The problem today is not so much one of de-
bating the role of reinforcement ad infinitum or determining the extinction curve in rats; rather, it is one of finding
out if it is really true, as it would seem, that apes can
communicate with man (Chapter 3).

Animals can at times return home from truly fantastic
distances. The important thing is not pursuit of the study
of tropisms, for the real question of orientation goes far
beyond that; rather, we must try to understand the mechanisms involved in the pigeon's return to its loft and the
salmon's return to its spawning grounds, bearing in mind
that in each instance the animal utilizes *a number of stimuli* which it integrates in various ways depending on the
circumstances. Though frequently attempted, it is futile to
try to assign all the responsibility for orientation to a single
stimulus (Chapter 4).

One of the vital activities of animals is reproduction.
The methods of sexual approach used by animals are innumerable; releasers constitute only one factor in sexual
relations and must be considered in combination with individual preference, hierarchical rank, etc. Sexual behavior
is a particularly fertile area for comparing animals and man
(Chapter 5).

Animal affectivity is now being explored in terms of its
ontogenesis by studying both prenatal and postnatal stress.
Disturbances on the order of autism have been experimentally induced in monkeys by radical isolation of young
animals, and by the same method it has also been possible
to investigate disturbances of maternal behavior. What is
important here is the *potential breakthrough toward an experimental psychopathology*, which could hardly be based
solely on experiments involving the frustration of conditioned rats (Chapter 6).

Animals very rarely live in isolation, however; usually they form larger or smaller groups. Congregation of individuals gives rise to various psychophysiological phenomena. One of them is group effect. Other such phenomena are hierarchies and complex types of organization in which one can even find examples of tradition and invention (Chapter 7).

The question of animal communication is currently being explored by a host of biologists using a variety of methods. Of special import is the systematic study of *communication in wild monkeys*, which is often disconcertingly reminiscent of human communication, and related research is seeking to determine the *role of tradition and innovation in simian societies*, which may one day help us to understand better how man developed. Similar efforts are underway to obtain more precise data on bird song, analyze the forms of communication used by bees and ants, etc. (Chapter 8).

Lastly, ethological definitions must themselves be revised and purged of the purely verbal or circular explanations that are still employed too often. We must re-examine the concepts of drive, motivation, and conflict, abandon the search for a general theory of learning, define critical periods more precisely, determine what is innate and what is acquired, and, at the same time, refrain from dogmatism (Chapter 9).

Above all, a simplistic or reductionist approach must be avoided. Ethology must be accepted for what it is, a science only recently come into being. Still very much in its infancy, ethology is far too young to indulge in overambitious general theories.

REFERENCES

Adler, H. E. (1963), Sensory factors in migration. *Anim. Behav.*, 2:566-577.
Ahrens, R. (1954), Beiträge zur Entwicklung des Physiognomie und Mimik-erkennens. *Z. Exp. Angew. Psychol.*, 2:402-454; 599-633.
Altmann, S. A. (1965), Stochastics of social communication. *J. Theoret. Biol.*, 8:490-522.
Anthouard, F. (1969), Observations préliminaires sur le comportement du rat blanc en semi liberté. *Rev. Compar. Anim.*, 3:25-38.
Archer, J. E. & Blackmann, D. E. (1971), Prenatal psychological stress and offspring behavior in rats and mice. *Develop. Psychol.*, 4:193-248.
Baerends, G. P. (1957), The ethological analysis of fish behavior. In: *Physiology of Fishes*, ed. Brown. New York: Academic Press.
Beach, F. A. (1961), *Hormones and Behavior.* New York: Cooper Square Publications.
———— (1965), *Sex and Behavior.* New York: Wiley.
Becker, G. (1971), Magnetfeld Einfluss auf die Galeriebau: Richtung bei Termiten. *Naturwiss.*, 58:60.
Berkowitz, L. (1963), *Aggression.* New York: McGraw-Hill.
Bernstein, A. & Roberts, M. de V. (1958), Computer vs. chess-player. *Sci. Amer.*, 198:96-105.
———— et al. (1958), A chess-playing program for the IBM 704 computer. *Proceedings of the Western Joint Computer Conference*, pp. 157-159.
Bingham, H. C. (1928), Sex development in apes. *Compar. Psychol. Monogr.*, 5.
Birukow, G. (1964), Aktivität und Orientierungsrythmik beim Kornkäfer *Calandra granaria. Z. Tierpsychol.*, 21:279-301.
Blauwelt, H. & Richmond, J. B. (1960), The development of contact between mother and offspring in ungulates. *Bull. Ecol. Soc. Amer.*, 41:91.
Bovet, J. (1960), Experimentelle Untersuchungen über das Heimfindevermögen von Mäuden. *Z. Tierpsychol.*, 17:728-755.
Bowlby, J. (1958), The nature of the child's tie to his mother. *Internat. J. Psycho-Anal.*, 39:350-373.
Brosset, A. (1973), Etude comparative de l'éthogenèse des comportements chez les rapaces Accipitridés et Falconidés. *Z. Tierpsychol.*, 32:386-417.
Brown, F. A., Barnwell, F. H., & Webb, H. M. (1964), Adaptation of the magnetoreceptive mechanism of mud snails to geomagnetic strength. *Biol. Bull.*, 127:221-231.
Bruce, H. M. (1961), Time relations in the pregnancy block induced in mice by strange males. *J. Reprod. Fertil.*, 2:138.
Brückman, D. (1962), Das Problem des Schweresinnes bei den Insekten. *Naturwiss.*, 49:28-33.

231

Brückner, G. H. (1933), Untersuchungen zur Tiersoziologie, insbesondere zur Auflösung der Familie. *Z. Psychol.*, 12:16-20.
Cairns, R. B. (1966), Attachment behavior of mammals. *Psychol. Rev.*, 73: 409-426.
Campbell, B. A. & Sheffield, F. O. (1953), Relation of random activity to food deprivation. *J. Comp. Physiol. Psychol.*, 46:320-322.
Carmichael, L. (1926-1927), The development of behavior in vertebrates experimentally removed from the influence of external stimulation. *Psychol. Rev.*, 33:51-58; 34:34-47.
Carpenter, C. R. (1934), A field study of the behavior and social relations of howling monkeys *Alouatta paliatta*. *Comp. Psychol. Monog.*, 10(2).
—— (1940), A field study in Siam on the behavior and social relations of the gibbon. *Comp. Psychol. Monogr.*, 16:1-212.
—— (1942), Social behavior in free ranging rhesus monkeys. *J. Comp. Psychol.*, 33:113-162.
Chauvin, R. (1956), L'animal en tant que constructeur; son intérêt pour la psychologie. *J. Psychol. Norm. Pathol.*, 1956:487-501.
—— (1958), Le comportement de construction chez *Formica rufa*. *Ins. Soc.*, 5:273-286.
—— (1959), La construction du dôme de *Formica rufa*. *Ins. Soc.*, 6:307-311.
—— (1960), Facteurs d'asymétrie et facteurs de régulation dans la construction du dôme chez *Formica rufa*. *Ins. Soc.*, 7:201-205.
—— (1965), Comportement des fourmis devant un obstacle "infranchissable." *Ins. Soc.*, 12:59-62.
—— (1968a), *Précis de Psychophysiologie, II. Le Comportement Animal.* Paris: Masson.
—— (1968b), *Traité de Biologie de l'Abeille.* Paris: Masson.
—— (1970), La technique de remplissage des objets creux chez *Formica polyctena*, stimuli déterminants. *Rev. Comp. Anim.*, 4:27-34.
—— (1971), Les lois de l'ergonomie chez les fourmis au cours du transport d'objets. *C. R. Acad. Sci.*, 273:1862-1865.
—— (1973), *Le Comportement Social chez les Animaux*, 2nd ed. Paris: Presses Universitaires de France.
Chauvin-Muckensturm, B. (1973), Solution brusque d'un problème nouveau chez le Pic épeiche. *Rev. Comp. Anim.*, 7:163-168.
Chomsky, N. (1965), *Aspects of the Theory of Syntax.* Cambridge: M.I.T. Press.
Christian, J. J. (1959), The role of endocrine and behavioral factors in the growth of mammalian populations. In: *Comparative Endocrinology.* New York: Gorbman & Wiley, pp. 71-120.
Collias, N. E. & Collias, E. C. (1962), An experimental study of the mechanism of nest building in a weaverbird. *Auk*, 79:568-595.
—— & Joos, M. (1953), The spectrographic analysis of sound signals of the domestic fowls. *Behaviour*, 5:175-188.
Coulson, J. C. (1968), Differences in the quality of birds resting in the centre and in the edges of a colony. *Nature*, 217:478-479.
Cowles, J. T. (1937), Food tokens as incentives for learning by chimpanzees. *Comp. Psychol. Monogr.*, 14:1-96.

Crawford, M. P. (1937), The cooperative solving of problems by young chimpanzees. *Comp. Psychol. Monogr.*, 14(2).

Crook, J. H. (1964), Field experiments on the nest construction and repair behavior of certain weaverbirds. *Proc. Zool. Soc. Lond.*, 142:217-255.

Cullen, J. M. (1972), Some principles of animal communication. In: *Non-verbal Communication*, ed. R. A. Hinde. London: Cambridge University Press, pp. 101-125.

Darchen, R. (1959), Les techniques de construction chez *Apis mellifica*. *Ann. Sci. Nat. Zool.*, 12 s.

Dechambre, R.-P. (1970), Effet de groupe et évolution des tumeurs ascitiques chez la souris. Thèse, Paris.

———— & Gosse, C. (1971), Effet de groupe et tumeurs greffés chez la souris. *Rev. Comp. Anim.*, 5:163-188.

Deleurance, E. P. (1957), Contribution à l'étude biologique des Polistes. I. L'activité de construction. *Ann. Sci. Nat. Zool.*, 19:99-222.

Delgado, J. M. R. (1967), Aggression and defense under cerebral radiocontrol. In: *Aggression and Defense*, ed. Clements & Lindsley. University of California Press.

Dembowski, J. (1933), Über die Plastizität der tierischen Handlungen. Beobachtungen und Versuche an Molanna Larven. *Zool. Jahrb.*, *Abt. Allg. Zool.*, 53:261-311.

Denenberg, V. H. (1962), An attempt to isolate critical periods of development in the rat. *J. Comp. Physiol. Psychol.*, 55:813-815.

———— & Bell, R. W. (1960), Critical periods for the effects of infantile experience on adult learning. *Science*, 131:227-228.

Denis, C. (1966-1967), Contribution à l'étude du comportement constructeur des larves de Trichoptères et problèmes relatifs à l'édification du fourreau. *Rev. Comp. Anim.*

DeVore, I., ed. (1965), *Primate Behavior*. New York: Holt, Rinehart & Winston.

Eibl-Eibesfeldt, I. (1970), *Ethology: The Biology of Behavior*. New York: Holt, Rinehart & Winston.

Emlen, S. T. (1969), Bird migration: Influence of physiological state upon celestial orientation. *Science*, 165:664-672.

Farkas, S. R. & Shorey, H. H. (1972), Chemical trail followed by flying insects. *Science*, 178:67-68.

Feigenbaum, E. A. (1963), The simulation of verbal learning behavior. In: *Computers and Thought: A Collection of Articles*, ed. E. A. Feigenbaum & J. Feldman. New York: McGraw-Hill, pp. 297-309.

———— & Feldman, J., eds. (1963), *Computers and Thought: A Collection of Articles*. New York: McGraw-Hill.

Ferster, A. (1964), Arithmetic behavior in chimpanzees. *Sci. Amer.*, 210:98-106.

Fisher, A. E. (1955), The effects of differential early treatment on the social and exploratory behavior of puppies. Unpublished doctoral dissertation, Pennsyvania State University.

Fries, J. C. de & Weir, M. C. (1964), Open field behavior of C 57 B1/6 J mice as a function of age, experience and prenatal maternal stress. *Psychonom. Sci.*, 1:389-390.

Frisch, K. von & Lindauer, M. (1954), Über die "Missweisung" bei der Richtungsanweisenden Tänzen der Bienen. *Naturwiss.*, 48:585-594.

Frith, H. J. (1962), *The Mallee Fowl*. Amgus & Robertson.

Gallais-Hamonno, F. & Chauvin, R. (1972), Simulation sur ordinateur de la construction du dôme et du ramassage des brindilles chez une Fourmi *Formica polyctena*. *C. R. Acad. Sci.*, 275:1275-1278.

Gallup, C. G. (1970), Chimpanzees: Self-recognition. *Science*, 167:86-87.

Gardner, R. A. & Gardner, B. T. (1969), Teaching sign language to a chimpanzee. *Science*, 165:664-672.

Geisler, M. (1961), Untersuchungen zur Tagesperiodik des Mistkäfers Geotrupes sylvaticus. *Z. Tierpsychol.*, 18:390-420.

Gonzales, R. C., Holmes, N. K., & Bitterman, M. E. (1967), Reversal learning and forgetting in bird and fish. *Science*, pp. 519-521.

Gottlieb, G. G. (1968), Imprinting in relation to parental and species identification by avian neonates. *J. Comp. Physiol. Psychol.*, 59:345-356.

Griffin, D. R. (1953), Sensory physiology and the orientation of animals. *Amer. Sci.*, 41:209-244.

—— & Hock, R. J. (1949), Airplane observations of homing birds. *Ecology*, 30:176-198.

Grohmann, J. (1939), Modifikation oder Funktionsreifung. *Z. Tierpsychol.*, 2:132-144.

Gromysz, K. (1960), Research on the plasticity of building behavior in caterpillars of the Bagworm *Psyche viciella*. *Fol. Biol.*, 8:351-416.

Gullahorn, J. T. & Gullahorn, J. E. (1963), A computer model of elementary social behavior. In: *Computers and Thought: A Collection of Articles*, ed. E. A. Feigenbaum & J. Feldman. New York: McGraw-Hill, pp. 375-386.

Guthrie, R. D. (1971), A new theory of mammalian rump patch evolution. *Behaviour*, 38:132-145.

Gwinner, E. (1961), Über die Entstachelungshandlung der Neuntötes. *Vogelwarte*, 21:36-47.

Hamburger, V. (1963), Some aspects of the embryology of behavior. *Quart. Rev. Biol.*, 38:349-352.

Hanna, H. M. (1961), The selection of case building materials by larvae of caddisflies (Trichoptera). *Proc. R. Entom. Soc. Lond.*, A 36:37-47.

Hansell, M. H. (1968), The house building behavior of the caddisfly larva *Silo pallipes* Fabricius: II. Description and analysis of the selection of small particles. III. The selection of large particles. *Anim. Behav.*, 16:562-584.

Harlow, H. F. (1958), The nature of love. *Amer. Psychol.*, 13:673-685.

—— (1962), Development of the second and third affectional system in macaque monkeys. In: *Research Approaches to Psychiatric Problems*, ed. Tourlentes et al.

—— & Harlow, M. K. (1965), The affectional systems. In: *Behavior of Nonhuman Primates*, Vol. 2, ed. A. M. Schreier, H. F. Harlow, & F. Stollnitz. New York: Academic Press, pp. 287-334.

—— & Warren, J. M. (1952), Formation and transfer of discrimination learning sets. *J. Comp. Physiol. Psychol.*, 42:482-489.

Harrisson, M. L. (1938), Visual stimulation and ovulation in pigeons. *Proc. R. Soc. B.*, 126.

Hasler, A. D. (1960), Homing orientation in migrating fishes. *Erg. Biol.*, 23:94-115.

Hayes, K. J. & Hayes, C. (1952), The intellectual development of a home raised chimpanzee. *Proc. Amer. Phil. Soc.*, 95:105-109.

Hebb, D. D. & Thompson, W. A. (1954), The social significance of animal studies. In: *Handbook of Social Psychology*, ed. G. Lindzey. Reading, Mass.: Addison-Wesley.

Heinroth, O. (1911) Beiträge zur Biologie namentlich Ethologie und Psychologie der Anatiden. *Verh. 5: Internat. Kong. Ornithol.*, pp. 589-702.

Herrnstein, R. J. & Loveland, D. H. (1964), Complex visual concepts in the pigeon. *Science*, 146:549-551.

Hess, E. H. (1965), Attitude and pupil size. *Sci. Amer.*, 212:46-54.

Hestermann, E. R. & Mykytowycz, R. (1968), Some observations on the odors of anal glands secretions from the rabbit Oryctolagus cuniculus. *CSIRO Wildl. Res.*, 13:71-81.

Hinde, R. A. (1970), *Animal Behavior: A Synthesis of Ethology and Comparative Psychology*, 2nd ed. New York: McGraw-Hill.

――― & Stevenson-Hinde, J. (1973), *Constraints on Learning; Limitations and Predispositions*. New York: Academic Press.

Hogan, J. A. (1965), An experimental study of conflict and fear: An analysis of behaviour of young chicks towards a mealworm. *Behaviour*, 25:45-97.

Homans, G. C. (1961), *Social Behavior: Its Elementary Forms*. New York: Harcourt, Brace & World.

Hull, C. L. (1952), *A Behavior System*. New Haven: Yale University Press.

Hunsaker, D. (1962), Ethological isolating mechanism in the Sceloporus torquatus group of lizards. *Evolution*, 16:62-74.

Hunt, J. McV. (1941), The effects of infant feeding frustration upon hoarding in the albino rat. *J. Abnorm. Soc. Psychol.*, 36:338-360.

Hutchison, R. E., Stevenson, J. G., & Thorpe, W. H. (1968), The basis for individual recognition by voice in the Sandwich tern (*Sterna sandwicensis*). *Behaviour*, 32:150-157.

Imanishi, K. (1957), Social behavior in Japanese monkeys *Macaca fuscata*. *Psychologia*, 1:47.

Itani, J. (1959), Paternal care in the wild Japanese monkey *Macaca fuscata*. *J. Primatol.*, 2:61.

Jaisson, P. (1974), Etude du développement de soins aux cocons chez la jeune Fourmi rousse *Formica polyctena* élévée en milieu précoce hétérospécifique. *C. R. Acad. Sci.*, 279:1205-1207.

Jay, P. (1968), *Primates: Studies in Adaptation and Variability*. New York: Holt, Rinehart & Winston.

Jolly, A. (1966), *Lemur Behavior: A Madagascar Field Study*. Chicago: University of Chicago Press.

Kalmus, H. (1955), The discrimination by the nose of the dog of individual human odours in particular of the odour of twins. *Brit. J. Anim. Behav.*, 3:25-31.

Kawamura, S. (1959), The process of subculture propagation among Japanese macaques. *J. Primatol.*, 2:43.

Kenyon, K. W. & Rice, D. W. (1958), Homing of Laysan albatrosses. *Condor*, 60:3-6.

Kister, J., Stein, P., Ulam, S., Walden, W., & Wells, M. (1957), Experiments in chess. *J. Assn. Computing Machinery*, April, 4(2):174-177.

Klüver, H. (1937), Reexamination of implement-using behavior in a *Cebus* monkey after an interval of 3 years. *Acta Psychol.*, 2:347-397.

Koch, S. (1954), In: *Modern Learning Theory*, ed. C. L. Hull. New York: Appleton-Century-Crofts.

Koehler, O. (1954), "Zählende" Vögel und vergleichende Verhaltensforschung. *Acta XI Cong. Internat. Ornith.*, pp. 588-598.

Köhler, W. (1921), *The Mentality of Apes*. New York: Random House, 1927.

Kogan, A. B. et al. (1966), *Proc. Conf. Effects of Magnetic Fields on Biological Objects*. Moscow, p. 37.

Kohts, N. (1935), Infant ape and human child. *Sci. Mem. Mus. Darwinianum*, 3.

Konishi, M. (1963), The role of auditory feedback in the vocal behaviour of the domestic fowl. *Z. Tierpsychol.*, 20:349-367.

Kortlandt, A. (1967), Handgebrauch bei freilebenden Schimpansen. In: *Handgebrauch und Verständigung bei Affen und Frühmenschen*, ed. Rensch. Bern: Huber, pp. 59-102.

Kramer, G. (1949), Long distance orientation. In: *Biology and Comparative Physiology of Birds*, Vol. 2, ed. A. J. Marshall. New York: Academic Press, pp. 341-371, 1961.

Krüger, — (1951), Über die Bahnflüge der Männchen der Gattungen Bombus und Psithyrus. *Z. Tierpsychol.*, 8:61-75.

Kuo, Z. Y. (1932), Ontogeny of embryonic behavior in Aves. IV. The influence of embryonic movement upon the behavior after hatching. *J. Comp. Physiol.*, 14:109-122.

Lagerspetz, K. (1964), Studies on the aggressive behavior of mice. *Suomi Tiedeakad. Toimituksia*, B. 131. Helsinki.

Lauer, J. & Lindauer, M. (1973), Die Beteiligung von Lernprozessen bei der Orientierung. *Fortschr. Zool.*, 21:349-370.

Lawick-Goodall, J. van (1965), Chimpanzee of the Gombe stream reserve. In: *Primate Behavior*, ed. I. DeVore. New York: Holt, Rinehart & Winston, pp. 425-473.

——— (1971), *In the Shadow of Man*. Boston: Houghton Mifflin.

Lee, S. van der & Boot, L. M. (1956), Spontaneous pseudopregnancy in mice II. *Acta Physiol. Pharmacol. Neerl.*, 5:213-214.

Lehr, E. (1967), *Z. Tierpsychol.*, 24:208-244.

Lehrman, D. S. (1953), A critique of Konrad Lorenz's theory of instinctive behavior. *Quart. Rev. Biol.*, 28:337-363.

Lenneberg, E. H. (1964), A biological perspective of language. In: *New Directions in the Study of Language*, ed. E. H. Lenneberg. Cambridge: M.I.T. Press.

Lethmate, J. & Dücker, G. (1973), Untersuchungen zum Selbsterkennen im

Spiegel bei Orang Utans und einigen anderen Affenarten. Z. Tierpsychol., 33:248-269.

Levine, S. & Alpert, M. (1959), Differential maturation of the central nervous system as a function of early experience. AMA Arch. Gen. Psychiat., 1:403-405.

——— & Lewis, G. W. (1958), Differential maturation of an adrenal response to cold stress in rats manipulated in infancy. J. Comp. Physiol. Psychol., 54:774-777.

Liddell, H. (1954), Conditioning and emotions. Sci. Amer., 190:48-57.

Lill, A. & Wood-Gush, D. G. M. (1965), Potential ethological isolating mechanisms and assortative mating in the domestic fowl. Behaviour, 26:16.

Lindauer, M. (1971), Orientierung der Bienen, neue Erkenntnisse, neue Rätsel. Rhein. Westf. Akad. Wiss. Vort., p. 218.

——— & Nedel, O. (1959), Ein Schweresinnesorgan der Honigbiene. Z. Vergl. Physiol., 41:405-434.

Lissmann, H. W. (1958), On the function and evolution of electric organs in fish. J. Exp. Biol., 35:156.

——— & Machin, K. E. (1958), The mechanism of object location in Gymnarchus niloticus and similar fish. J. Exp. Biol., 35:451.

Logan, F. A. (1959), The Hull Spence approach. In: Psychology: A Study of a Science, ed. S. Koch. New York: McGraw-Hill.

Loizos, C. (1967), Play behaviour in higher primates. In: Primate Ethology, ed. Morris. London: Weidenfeld and Nicolson, pp. 176-218.

Lorenz, K. (1952), Die Entwicklung der vergleichenden Verhaltensforschung in den letzten zwölf Jahren. Verh. Dtsch. Zool. Ges. Freiburg, pp. 36-58.

——— (1963), On Aggression. New York: Bantam Books.

——— (1965), Evolution and Modification of Behavior. Chicago: University of Chicago Press.

——— (1966), Stammes- und Kulturgeschichtliche Ritenbildung. Mitt. Max Planck Inst. Ges., 1:3-30.

Markl, H. (1962), Borstenfelder an den Gelenken als Schweresinnesorganes bei Ameisen und anderen Hymenopteren. Z. Vergl. Physiol., 45:475-569.

Marler, P. (1965), Communication in monkeys and apes. In: Primate Behavior, ed. I. DeVore. New York: Holt, Rinehart & Winston, pp. 544-584.

Maslow, A. H. (1940), Dominance quality and social behavior in infrahuman primates. J. Soc. Psychol., 11:313-324.

Mason, J. W. (1959), Psychological influences on the pituitary-adrenal cortical system. Recent Progress in Hormone Research, 15:345-389.

——— & Brady, J. V. (1964), The sensitivity of psychoendocrine systems to social and physical environment. In: Psychobiological Approaches to Social Behavior, ed. P. H. Leiderman & D. Shapiro. Stanford: Stanford University Press, pp. 4-23.

Matthews, G. V. T. (1955), Bird Navigation, 2nd ed. London: Cambridge University Press, 1968.

McClintock, M. K. (1971), Menstrual synchrony and suppression. Nature, 229:244-245.

Messmer, E. & Messmer, I. (1956), Die Entwicklung der Lautäusserungen und einiger Verhaltensweisen der Amsel Turdus merula merula unter natürliche Bedingungen und nach Einzelaufzucht inschalldichten Räumen. *Z. Tierpsychol.*, 13:341-441.

Meyer, D. R. (1952), The stability of human gustatory sensibility during changes in time of food deprivation. *J. Comp. Physiol. Psychol.*, 45:373-376.

Miller, R. E., Banks, J. H., & Ogawa, N. (1962), Communication of affect in cooperative conditioning of Rhesus monkeys. *J. Abnorm. Soc. Psychol.*, 64:343-348.

Morris, D. (1956), The feather posture of birds and the problem of the origin of social signals. *Behaviour*, 9:75-113.

Muckensturm, B. (1965a), Possibilités inattendues de manipulation chez l'Epinoche. *C. R. Acad. Sci.*, 260:3183-3184.

——— (1965b), Le nid et le territoire chez l'Epinoche *Gasterosteus aculeatus. C. R. Acad. Sci.*, 260:4825.

——— (1969), La signification de la livrée nuptiale de l'Epinoche. *Rev. Comp. Anim.*, 3(3):39-65.

Müller-Velten, H. (1966), Über den Angstgeruch bei der Hausmaus. *Z. Vergl. Physiol.*, 52:401-429.

Newell, A., Shaw, J. C., & Simon, H. A. (1963), Chess-playing programs and the problem of complexity. In: *Computers and Thought: A Collection of Articles*, ed. E. A. Feigenbaum & J. Feldman. New York: McGraw-Hill, pp. 39-70.

Nisbet, I. C. T. & Drury, W. H. (1967), *Bird Banding*, 38:173-186.

Novick, A. (1959), Acoustic orientation in the cave swiftlet. *Biol. Bull.*, 117: 497-503.

Pallaud, B. (1972), L'apprentissage par observation chez les Rongeurs. *Colloques Internat. C.N.R.S.*, No. 198:265-285.

Pardi, L. (1960), Innate components in the solar orientation of littoral Amphipods. *Cold Spring Harb. Sympos. Quantit. Biol.*, 25:395-401.

Payne, R. S. & McVay, S. (1971), Songs of humpback whales. *Science*, 173:585-597.

Perdeck, A. C. (1958), Two types of orientation in migrating starlings *Sturnus vulgaris* and chaffinches *Fringilla coelebs*, as revealed by displacement experiments. *Ardea*, 46:1-37.

Piaget, J. (1936), *The Origins of Intelligence in Children.* New York: International Universities Press, 1952.

Ploog, D. & Melnechuk, T. (1971), Are apes capable of language? *Nevrosci. Res. Prog. Bull.*, 9:599-700.

Premack, D. (1970), The education of Sarah. *Psychol. Today*, 4, Sept.: 54-58.

Preyer, W. (1885), *Spezielle Physiologie des Embryo.* Leipzig: Grieben.

Rabaud, E. (1929), *Phénomène social et sociétés animales.* Paris: Alcan.

Razran, G. (1961), Observable unconscious and inferable conscious. *Psychol. Rep.*, 68:81-147.

Regen, J. (1924), Über die Orientierung des Grillenweibchens nach dem Stidulationschall des Männchens. *Sitz. Ber. Akad. Wiss. Wien, Math. Nat. Kl.*, p. 132.

Rensch, B. (1965), Die höchsten Lernleistungen der Tiere. *Naturwiss. Rdsch.*, 18:91-101.

—— (1973), *Gedächtnis, Begriffsbildung und Planhandlungen bei Tieren.* Berlin: Parey.

—— & Altevogt, B. (1953), Visuelles Lernvermögen eines indischen Elefanten. *Z. Tierpsychol.*, 10:119-134.

—— & Dücker, G. (1959), Versuche über visuelle Generalisation bei einer Schleichkatze. *Z. Tierpsychol.*, 16:671-692.

Richard, B. (1964), Les matériaux de construction du Castor leur signification pour ce rongeur. *Z. Tierpsychol.*, 21:592-601.

Riesen, A. H., Greenberg, A., Granston, A. & Fantz, R. L. (1953), Solution of patterned string problems by young gorillas. *J. Comp. Physiol. Psychol.*, 46:19-22.

Roeder, K. D. (1964), Aspects of the noctuids tympanic nerve response having significance in the avoidance of bats. *J. Ins. Physiol.*, 10:529-546.

Ropartz, P. (1967), *Etude du Déterminisme olfactif de l'Effet de Groupe chez la Souris.* Thèse d'Etat, Strasbourg.

Ross, H. H. (1964), Evolution of caddisworm cases and nets. *Amer. Zool.*, 4:209-220.

Rowell, T. E. & Hinde, R. A. (1962), Vocal communication by the rhesus monkeys *Macaca mulatta. Proc. Zool. Soc. Lond.*, 138:279-294.

Rumbaugh, D. M., Gill, T. V., & Glaserfeld, E. C. von (1973), Reading and sentence completion by a chimpanzee (Pan). *Science*, 182:731-733.

Sackett, G. P. (1970), Isolation rearing in monkeys. Diffuse and specific effects on later behaviour. In: *Modèles animaux du Comportement humain*, ed. R. Chauvin. Paris, pp. 60-110.

Sade, D. S. (1967), Determinants of dominance in a group of free ranging Rhesus monkeys. In: *Social Communication among Primates*, ed. S. A. Altmann. Chicago: University of Chicago Press, pp. 99-114.

Saila, S. B. & Shappy, R. A. (1963), Random movements and orientation in salmon migration. *J. Cons. Internat. Explor. de la Mer*, 28:154-166, 440-443.

Samuel, A. L. (1963), Some studies in machine learning using the game of checkers. In: *Computers and Thought: A Collection of Articles*, ed. E. A. Feigenbaum & J. Feldman. New York: McGraw-Hill, pp. 71-108.

Sauer, F. (1961), Further studies on the nocturnal orientation of nocturnally migrating birds. *Psychol. Forsch.*, 26:224-244.

Schaller, C. B. (1963), *The Mountain Gorilla: Ecology and Behavior.* Chicago: University of Chicago Press.

Schiller, P. H. (1949), Innate motor action as a basis of learning: Manipulative patterns in the chimpanzee. In: *Instinctive Behavior*, ed. C. Schiller. New York: International Universities Press, 1957.

Schleidt, W. M., Schleidt, M., & Magg, M. (1960), Störungen der Mutter-Kind Beziehung bei Truthühnern durch Gehörverlust. *Behaviour*, 16:254-260.

Schneider, F. (1961), Beeinflussung der Aktivität des Maikäfers durch Veränderung der gegenseitigen Lage magnetischer und elektrischer Felder. *Mitt. Schweiz. Ent. Ges.*, 33:223-237.

——— (1963), Ultraoptische Orientierung des Maikäfers *Melolontha vulgaris* in künstlichen elektrischen und magnetischen Feldern. *Erg. Biol.,* 26:147-157.

Schneirla, T. C. (1965), Aspects of stimulation and organization in approach-withdrawal processes underlying vertebrate behavioral development. In: *Advances in the Study of Behavior,* ed. D. S. Lehrman et al. New York: Academic Press.

Schöne, H. (1959), Die Lageorientierung mit Statolithenorganen und Augen. *Ergeb. Biol.,* 21:161-209.

Schulte, E. H. (1970), Unterschiede im Lern- und Abstraktionsvermögen von binokular und monokular sehenden Hühnern. *Z. Tierpsychol.,* 27:946-970.

Scott, J. P. (1963), The process of primary socialization in canine and human infants. *Monog. Soc. Res. Child Develop.,* 28.

——— (1966), Agonistic behavior of mice and rats; A review. *Amer. Zool.,* 6:683-701.

Secord, P. F. (1958), Facial features and inference processes. In: *Person Perception and Intrapersonal Behavior,* ed. Tagiuri & Petrullo. Stanford: Stanford University Press, pp. 300-315.

Seitz, P. F. D. (1954), The effect of infantile experiences upon adult behavior in animal subjects. *Amer. J. Psychol.,* 110:916-927.

Selfridge, O. G. & Neisser, U. (1963), Pattern recognition by machine. In: *Computers and Thought: A Collection of Articles,* ed. E. A. Feigenbaum & J. Feldman. New York: McGraw-Hill, pp. 237-250.

Seligman, M. E. P. (1970), On the generality of the laws of learning. *Psychol. Rev.,* 77:406-418.

Sevenster, P. (1961), A causal analysis of a displacement activity. *Behaviour Suppl.,* 9:1-170.

Shipley, W. U. (1963), The demonstration in the domestic guinea pig of a process resembling classical imprinting. *Anim. Behav.,* 2:470-474.

Skinner, B. F. (1951) How to teach animals. *Sci. Amer.,* 191:262-269.

——— (1953), *Science and Human Behavior.* New York: Macmillan.

Sokolovsky, A. (1908), *Beobachtungen über die Psyche der Menschenaffen.* Neue Frankfurt Verlag.

Southern, W. E. (1970), *Amer. Zool.,* 10.

Staropolska, S. & Dembowski, J. (1958), An attempt of analyzing the variability in the behavior of the caddisfly larva *Molanna angustata. Acta Biol.,* 6:301-306.

Steiner, A. L. (1970), Etude descriptive de quelques activités et comportements de base du Spermophile *Spermophilus columbianus. Rev. Comp. Anim.,* 4:3-21; 23-42.

Struhsaker, T. T. (1967), Behavior of vervet monkeys *Cercopithecus aethiops. Univ. Calif. Publ. Zool.,* 82:1-64.

Sudd, J. H. (1966), *An Introduction to the Study of Ant Behavior.* London: Arnold.

Suomi, S. T. (1972), Social development of rhesus monkeys reared in an enriched laboratory environment. Paper presented at the 20th International Congress of Psychology, Tokyo, August.

Thorpe, W. H. (1975), *Animal Nature and Human Nature*. New York: Doubleday.

Tinbergen, N. (1951), *The Study of Instinct*. London: Oxford University Press.

—— (1959), Comparative studies of the behavior of gulls (*Laridae*), a progress report. *Behaviour*, 15:1-70.

Tinklepaugh, O. L. (1932), Multiple delayed reactions with chimpanzees and monkeys. *J. Comp. Psychol.*, 13:200-243.

Turing, A. M. (1950), Computing machinery and intelligence. *Mind*, October, 59(n.s. 236):433-460. Reprinted in: *The World of Mathematics*, Vol. 4, ed. J. R. Newman. New York: Simon & Schuster, 1954.

Washburn, S. L. & DeVore, I. (1961), The social life of baboons. *Sci. Amer.*, 204:62.

Whitten, W. K. (1959), Occurrence of anoestrus in mice caged in groups. *J. Endocrin.*, 18:107.

Wiepkema, P. R. (1961), An ethological analysis of the reproductive behaviour of the bitterling. *Arch. Néer. Zool.*, 14:103-199.

Wilson, A. P. (1968), Social behavior of free ranging monkeys with an emphasis on aggression. Unpublished doctoral dissertation, University of California, Berkeley.

Wiltschko, W. (1968), Über den Einfluss statischer Magnetfelder auf die Zugorientierung der Rotkehlchen *Erithacus rubecula*. *Z. Tierpsychol.*, 25:537-558.

Wolfe, J. B. (1936), Effectiveness of token rewards for chimpanzees. *Comp. Psychol. Monogr.*, 12:1-72.

Woods, P. J., Ruckelshaus, S. I., & Bowling, D. M. (1960), Some effects of "free" and "restricted" environmental rearing conditions upon adult behavior. *Therap. Psychol. Rep.*, 6:191-200.

Wright, R. H. (1963), Molecular vibrations and insect attractions. *Nature*, 198:455-459.

Wymann, R. L. & Ward, J. D. (1973), The development of behavior in the cichlid fish *Etroplus maculatus B*. *Z. Tierpsychol.*, 33:461-491.

Yamada, M. A. (1963), A study of blood relationship in the natural society of the Japanese macaque. *Primates*, 4:43-65.

Yerkes, R. M. (1948), *Chimpanzees: A Laboratory Colony*, 4th ed. New Haven: Yale University Press.

—— & Elder, J. H. (1936), Oestrus, receptivity and mating in chimpanzees. *Comp. Psychol. Monogr.*, 13:1-39.

Zuckerman, S. (1932), *The Social Life of Monkeys and Apes*. London.

INDEX